Methods in Molecular Biology

For further volumes:
http://www.springer.com/series/7651

For over 35 years, biological scientists have come to rely on the research protocols and methodologies in the critically acclaimed *Methods in Molecular Biology* series. The series was the first to introduce the step-by-step protocols approach that has become the standard in all biomedical protocol publishing. Each protocol is provided in readily-reproducible step-by-step fashion, opening with an introductory overview, a list of the materials and reagents needed to complete the experiment, and followed by a detailed procedure that is supported with a helpful notes section offering tips and tricks of the trade as well as troubleshooting advice. These hallmark features were introduced by series editor Dr. John Walker and constitute the key ingredient in each and every volume of the *Methods in Molecular Biology* series. Tested and trusted, comprehensive and reliable, all protocols from the series are indexed in PubMed.

Dynamics of Bacteria-Mucus Interactions

Edited by

Inka Brockhausen

Department of Biomedical and Molecular Sciences, Queen's University, Kingston, ON, Canada

Humana Press

Editor
Inka Brockhausen
Department of Biomedical
and Molecular Sciences
Queen's University
Kingston, ON, Canada

ISSN 1064-3745 ISSN 1940-6029 (electronic)
Methods in Molecular Biology
ISBN 978-1-0716-4626-7 ISBN 978-1-0716-4627-4 (eBook)
https://doi.org/10.1007/978-1-0716-4627-4

This Humana imprint is published by the registered company Springer Science+Business Media, LLC, part of Springer Nature.
The registered company address is: 1 New York Plaza, New York, NY 10004, U.S.A.

Preface

It is hoped that this book will alert people to the importance of bacteria in and around your body, which are harmless and in fact beneficial. In contrast, pathogenic bacteria require an understanding of how they operate and how we can protect ourselves from them.

This issue on Dynamics of mucus-bacteria interactions includes diverse protocols that form a basis for studies of bacterial characteristics and mutual and dynamic interactions with the host mucosa. In light of growing antibiotic resistance, research in this field is critical and timely. The interactions between bacteria and the host are multifaceted, extremely complex and dynamic, and essential for health and protection from pathogens.

Layers of mucus cover the gastrointestinal, airway, genital, and oral epithelia. The human gut has evolved to host trillions of microbes that are mainly harmless, commensal, probiotic bacteria. Bacteria utilize the mucin glycoproteins in mucus for adhesion and as nutrients through the degradation of oligosaccharides. Mucin glycoproteins are heterogeneous in structure and carry hundreds of simple and complex O-linked glycans with multitudes of functions. Mucins play important roles in the immune system, and they control the composition of the microbiome, impacting adhesion events and biofilm formation. Bacteria also produce highly complex polysaccharides that are involved in their survival. Knowledge of bacterial interactions with mucus can help to develop new antibacterial strategies.

Protocols are presented here to examine mucus-bacteria interactions using various model systems. These include rodent models, cultured cells, and in vitro studies that allow investigations of physiological consequences of bacterial products, infection dynamics, in vivo adhesion, and invasion as well as responses to pathogens. Mice can be used to examine infection and pathogenicity by intestinal pathogens, and test novel prophylactic or treatment methods.

Cultured epithelial cells serve as models to analyze molecular details of adhesion and invasion by specific bacteria. These cells can be useful to study growth and colonization of specific pathogens and the effect of co-culturing different bacteria. Infections of epithelial cells can be detected by immunostaining and imaging of cells. Protocols to prepare adhesion assays using cultured and polarized lung epithelial cells are a critical basis to study the infectious process.

The prevalence of bacterial antibiotic resistance is constantly increasing. The gastric pathogen *Helicobacter pylori* can be treated with antibiotics but additional strategies are required. Lactic acid bacteria could enhance effects of treatment and reduce *H. pylori* growth and virulence. Bacterial antibiotic resistance is often mediated by production of beta-lactamases that can degrade lactam-antibiotics such as penicillins. The development of novel beta-lactamase inhibitors is crucial to target opportunistic pathogens. Their susceptibility to novel inhibitors can be tested in bacteria cells with luminescent biosensors.

The opportunistic pathogen *Acinetobacter baumannii* is one of the most dangerous pathogens since it developed broad resistance to antibiotics. Biofilms of *A. baumannii* are found on epithelial cells and abiotic surfaces. Biofilms form a barrier against host and environmental impacts and have critical roles in the survival and pathogenicity of bacteria. Biofilms are composed of mucus components, bacterial polysaccharides, proteins, and DNA and often contain several different strains of bacteria. Antibiotics often cannot reach bacteria in biofilms, and analyses of the production and potential inhibition of biofilms are important to prevent infections. The motility and movements of bacteria are crucial factors in the dynamics of bacterial interactions with mucosa. Swimming patterns play an important role in the formation and properties of biofilms. Time-lapse imaging can track bacteria within mucus and their speed, random movements, and persistence can be quantified. This can also give valuable information on mucus permeability and may be applied to immune cell migration through mucus.

Interactions with mucins can be studied by labeling bacteria with biotin and visualizing and quantifying bacteria with slot-blots and imaging systems. Mucin degradation is a critical event in providing a nutrient-rich niche for bacteria. Slot-blot analyses can help to define the specific mucins and their glycans that are involved in adhesion and are degraded. The adhesion of diverse bacterial strains isolated from patients can be assessed using mucin-coated microplates and counting colony-forming units of adhering bacteria. Host secretory IgA and IgG function in controlling microbes, for example, in the oral cavity. Highly sensitive methods are required to analyze these interactions, and glycan arrays can assess the glycan specificity of interactions. To determine host-pathogen adhesion, imaging techniques such as confocal microscopy using immuno-staining and fluorescence can be applied to cultured cells after pathogen challenge. Similarly, using tissue sections from rodent models, the exact localization and dissemination of pathogens can be visualized.

The adhesion of bacteria to specific mucins can be revealed with sensitive methods such as infrared imaging. Mucins can also be used to block the adhesion of bacteria and provide information leading to reduction of infections. Mucins are difficult to analyze due to their large size and heterogeneity. Slot-blot methods can be used to immobilize mucins for lectin or antibody staining and evaluation of the glycans present after hydrolase treatments. Immobilized mucins also serve to study bacterial adhesion. Gastric mucins include MUC5AC in humans and other mammals. The interactions between *Helicobacter pylori* and this mucin can be shown by immunofluorescence staining of gastric epithelial cells from guinea pig. This will determine the colonization patterns of this pathogen.

Host-pathogen interactions can be evaluated by several methods, including confocal microscopy, immunostaining, fluorescence imaging, and measuring transepithelial resistance and 3D model construction of epithelial layers. This allows the properties of the epithelial barrier and the progress of infections to be determined.

The methods included here meant to allow studies of mucin degradation by specific hydrolases. Mucins are degraded by enzymes produced by the microbiome. Cleavage of sugars by glycosylhydrolases creates a "degradome" and exposes underlying glycan epitopes that represent another layer of bacteria-binding receptors.

A great number of different adhesion proteins and lectins are involved in the interactions between bacteria and mucus, representing another level of complexity. Mucins are the main adhesion receptors for bacteria in mucus. Thus, biochemical studies of the binding specificities of bacterial adhesins can help to predict the niches for specific bacteria and develop inhibitors of adhesion. The specificity of lectins can be studied using glycan-linked resins. The competitive release of bound lectins by simple glycans can show the relative binding affinity to various glycans. There are several databases that include adhesins, and bioinformatic tools allow the usage of a wealth of available data. They can help to estimate specificities of lectins toward mucin receptors. The protocol described here helps to navigate through these resources to identify lectins and their glycan recognition of mucus-associated bacteria.

Bacterial adherence to mucins can be quantified on microplates coated with various mucins. In chronic infections such as in the airways of cystic fibrosis, many clones of bacterial species exist that may exhibit differences in adherence properties. The mucin adhesion method can distinguish inter-clonal properties and also evaluate mucin-derived oligosaccharides as competitors of adhesion.

Lipoteichoic acids (LTA) are complex cell wall polysaccharides of Gram-positive bacteria that colonize mucus. The structures of LTA show slight variations between bacteria. Thus, methods to isolate and characterize these polysaccharides are a basis for vaccine development against Gram-positive pathogens.

The combined efforts of the authors in this book that are providing their established protocols are a solid basis to expand or enter the field of mucus-bacteria interactions by students, postdoctoral fellows, and researchers that are driven to discover how we can live with so many bacteria around us.

Kingston, ON, Canada ***Inka Brockhausen***

Contents

Contributors

LUIS A. ACTIS • *Department of Microbiology, Miami University, Oxford, OH, USA*
ARUN K. BHUNIA • *Molecular Food Microbiology Laboratory, Department of Food Science, Purdue University, West Lafayette, IN, USA; Purdue Institute of Inflammation, Immunology, and Infectious Disease, Purdue University, West Lafayette, IN, USA; Department of Comparative Pathobiology, Purdue University, West Lafayette, IN, USA*
SYLVIE BINDA • *Lallemand Health Solutions, Rosell Institute for Microbiome and Probiotics, Montreal, Quebec, Canada*
NICOLAI V. BOVIN • *Shemyakin-Ovchinnikov Institute of Bioorganic Chemistry, Russian Academy of Sciences, Moscow, Russia*
STÉPHANE BRONNER • *Lallemand Health Solutions, Rosell Institute for Microbiome and Probiotics, Montreal, Quebec, Canada*
DAELYN G. CARROLL • *Microbiology and Immunology, The University of Western Ontario, London, ON, Canada*
MAGDALENA CHMIELA • *Faculty of Biology and Environmental Protection, Department of Immunology and Infectious Biology, Institute of Microbiology, Biotechnology and Immunology, University of Lodz, Lodz, Poland*
ZINA CHOUIT • *Faculté des sciences et technologies, UMR 8576 CNRS, Unité de glycobiologie structurale et fonctionnelle, Villeneuve d'Ascq, France*
CAROLE CREUZENET • *Microbiology and Immunology, The University of Western Ontario, London, ON, Canada*
PETER L. DAVIES • *Department of Biomedical and Molecular Sciences, Queen's University, Kingston, ON, Canada*
KATYA DOUCHANT • *Department of Biomedical and Molecular Sciences, Queen's University, Kingston, ON, Canada; Gastrointestinal Disease Research Unit (GIDRU), Queen's University, Kingston, ON, Canada*
RISHI DROLIA • *Cellular and Molecular Microbiology Laboratory, Department of Biological Science, Old Dominion University, Norfolk, VA, USA; Center for Bioelectronics, Old Dominion University, Norfolk, VA, USA; Molecular Food Microbiology Laboratory, Department of Food Science, Purdue University, West Lafayette, IN, USA; Purdue Institute of Inflammation, Immunology, and Infectious Disease, Purdue University, West Lafayette, IN, USA*
ROBERT EVES • *Department of Biomedical and Molecular Sciences, Queen's University, Kingston, ON, Canada*
STEVEN E. FIESTER • *Department of Biological Sciences, Florida Gulf Coast University, Fort Myers, FL, USA; Department of Chemistry, Furman University, Greenville, SC, USA; School of Health Research, Clemson University, Clemson, SC, USA; Department of Pathology, Prisma Health Upstate, Greenville, SC, USA; College of Medicine, Family Medicine Residency at Lee Health, Florida State University, Fort Myers, FL, United States; College of Medicine, Internal Medicine Residency at Lee Health, Florida State University, Cape Coral, FL, United States*
JOLENE M. GARBER • *VaxAlta Inc. and Department of Medical Microbiology and Immunology, University of Alberta, Edmonton, AB, Canada*

WERONIKA GONCIARZ • *Faculty of Biology and Environmental Protection, Department of Immunology and Infectious Biology, Institute of Microbiology, Biotechnology and Immunology, University of Lodz, Lodz, Poland*

JILL GREENLAW • *Gastrointestinal Disease Research Unit (GIDRU), Queen's University, Kingston, ON, Canada; Translational Medicine, Queen's University, Kingston, ON, Canada*

YANN GUERARDEL • *Faculté des sciences et technologies, UMR 8576 CNRS, Unité de glycobiologie structurale et fonctionnelle, Villeneuve d'Ascq, France*

ANNE IMBERTY • *Université Grenoble Alpes, CNRS, CERMAV, Grenoble, France*

MITCHELL A. JEFFS • *Department of Biomedical and Molecular Sciences, Queen's University, Kingston, ON, Canada*

CHUNSHENG JIN • *Proteomics Core Facility at Sahlgrenska Academy, University of Gothenburg, Gothenburg, Sweden*

NICLAS G. KARLSSON • *Faculty of Health Sciences, Department of Life Sciences and Health, Oslo Metropolitan University, Oslo, Norway*

MARINA V. LAVRENTEVA • *Biotechnology Department, MIREA—Russian Technological University, Moscow, Russia*

RENAUD LÉONARD • *Faculté des sciences et technologies, UMR 8576 CNRS, Unité de glycobiologie structurale et fonctionnelle, Villeneuve d'Ascq, France*

SARA K. LINDÉN • *Department of Medical Biochemistry and Cell Biology, Institute of Biomedicine, Sahlgrenska Academy, University of Gothenburg, Gothenburg, Sweden*

FRÉDÉRIQUE LISACEK • *Swiss Institute of Bioinformatics, Geneva, Switzerland; Computer Science Department, UniGe, Geneva, Switzerland; Section of Biology, UniGe, Geneva, Switzerland*

CHRISTOPHER T. LOHANS • *Department of Biomedical and Molecular Sciences, Queen's University, Kingston, ON, Canada*

ANA S. LUIS • *Department of Medical Biochemistry, Institute for Biomedicine, Sahlgrenska Academy, University of Gothenburg, Gothenburg, Sweden*

MARISSA L. MACIEJ-HULME • *Faculty of Health Sciences, Department of Life Sciences and Health, Oslo Metropolitan University, Oslo, Norway*

EDWIGE MADEC • *Faculté des sciences et technologies, UMR 8576 CNRS, Unité de glycobiologie structurale et fonctionnelle, Villeneuve d'Ascq, France*

JEAN W. MARBOU TAKOUGOUM • *Microbiology and Immunology, The University of Western Ontario, London, ON, Canada*

MARTHE TOFTHAGEN • *Faculty of Health Sciences, Department of Life Sciences and Health, Oslo Metropolitan University, Oslo, Norway*

MAHI M. MOHIUDDIN • *Microbiology and Immunology, The University of Western Ontario, London, ON, Canada*

HARALD NOTHAFT • *VaxAlta Inc. and Department of Medical Microbiology and Immunology, University of Alberta, Edmonton, AB, Canada*

DAVID OHAYON • *Section of Biology, UniGe, Geneva, Switzerland*

EMILY J. OHNECK • *Medpace, Inc., Cincinnati, OH, USA*

HANNAH J. PARKS • *Infection Biology Group, Wellcome-Wolfson Institute for Experimental Medicine, Queen's University Belfast, Belfast, UK*

PIYUSH PATEL • *Division of Clinical Cancer Epidemiology, Department of Oncology, Institute of Clinical Sciences, The Sahlgrenska Academy, University of Gothenburg, Gothenburg, Sweden; Department of Infectious Diseases, Institute of Biomedicine, The Sahlgrenska Academy, University of Gothenburg, Gothenburg, Sweden*

WILLIAM F. PENWELL • *Department of Biology and Marine Science, Jacksonville University, Jacksonville, FL, USA*

MIA S. PETRUCCI • *Department of Biological Sciences, Florida Gulf Coast University, Fort Myers, FL, USA*

SVETLANA M. POLYAKOVA • *Shemyakin-Ovchinnikov Institute of Bioorganic Chemistry, Russian Academy of Sciences, Moscow, Russia*

MACARENA P. QUINTANA-HAYASHI • *Department of Medical Biochemistry and Cell Biology, Institute of Biomedicine, Sahlgrenska Academy, University of Gothenburg, Gothenburg, Sweden*

LUIS A. RAMOS • *Department of Biological Sciences, Florida Gulf Coast University, Fort Myers, FL, USA*

CATHERINE ROBBE-MASSELOT • *Faculté des sciences et technologies, UMR 8576 CNRS, Unité de glycobiologie structurale et fonctionnelle, Villeneuve d'Ascq, France*

YANNICK ROSSEZ • *Faculté des sciences et technologies, UMR 8576 CNRS, Unité de glycobiologie structurale et fonctionnelle, Villeneuve d'Ascq, France*

LICÍNIA SANTOS • *Department of Medical Biochemistry and Cell Biology, Institute of Biomedicine, Sahlgrenska Academy, University of Gothenburg, Gothenburg, Sweden*

BORIS SCHNIDER • *Swiss Institute of Bioinformatics, Geneva, Switzerland; Computer Science Department, UniGe, Geneva, Switzerland*

PRAMEET M. SHETTI • *Department of Biomedical and Molecular Sciences, Queen's University, Kingston, ON, Canada; Gastrointestinal Disease Research Unit (GIDRU), Queen's University, Kingston, ON, Canada; Translational Medicine, Queen's University, Kingston, ON, Canada; Pathology and Molecular Medicine, Queen's University, Kingston, ON, Canada*

NADEZHDA V. SHILOVA • *Shemyakin-Ovchinnikov Institute of Bioorganic Chemistry, Russian Academy of Sciences, Moscow, Russia*

FEI SJÖBERG • *Division of Clinical Cancer Epidemiology, Department of Oncology, Institute of Clinical Sciences, The Sahlgrenska Academy, University of Gothenburg, Gothenburg, Sweden; Department of Infectious Diseases, Institute of Biomedicine, The Sahlgrenska Academy, University of Gothenburg, Gothenburg, Sweden*

COREY A. STEVENS • *Department of Biological Engineering, Massachusetts Institute of Technology, Cambridge, MA, USA*

CHRISTINE M. SZYMANSKI • *VaxAlta Inc. and Department of Medical Microbiology and Immunology, University of Alberta, Edmonton, AB, Canada; Complex Carbohydrate Research Center and Department of Microbiology, University of Georgia, Athens, GA, USA*

BURKHARD TÜMMLER • *Department of Pediatric Pneumology, Allergology and Neonatology, Hannover Medical School, Hannover, Germany; Biomedical Research in Endstage and Obstructive Lung Disease Hannover (BREATH), German Center for Lung Research (DZL), Hannover Medical School, Hannover, Germany*

MIGUEL A. VALVANO • *Infection Biology Group, Wellcome-Wolfson Institute for Experimental Medicine, Queen's University Belfast, Belfast, UK*

OLIVIER VIDAL • *Faculté des sciences et technologies, UMR 8576 CNRS, Unité de glycobiologie structurale et fonctionnelle, Villeneuve d'Ascq, France*

CHRIS VIETS • *Department of Biological Engineering, Massachusetts Institute of Technology, Cambridge, MA, USA*

ELIZABETH C. WALLS • *Department of Biological Sciences, Florida Gulf Coast University, Fort Myers, FL, USA*
CORY Q. WENZEL • *VaxAlta Inc. and Department of Medical Microbiology and Immunology, University of Alberta, Edmonton, AB, Canada*
SAMANTHA A. WHITESIDE • *Microbiology and Immunology, The University of Western Ontario, London, ON, Canada*

Chapter 1

Breaking Barriers: A Protocol to Investigate Intestinal Pathogen Adhesion, Invasion, and Translocation Through In Vitro, In Vivo, and Imaging Technologies

Rishi Drolia and Arun K. Bhunia

Abstract

The gastrointestinal tract is a crucial battleground in microbial pathogenesis, where pathogens must adhere to, invade, and translocate through intestinal epithelial cells to establish infection and cause disease. Investigating these processes is critical for understanding the mechanisms of infection and host defense. In vitro assays using intestinal cell models enable precise analysis of microbial adhesion, invasion, and translocation at the cellular level. In contrast, in vivo, rodent models provide valuable insights into the complex interactions between pathogens, host tissues, and immune responses. Advanced imaging techniques, such as confocal microscopy, offer high-resolution visualization of host–pathogen interactions, allowing researchers to observe pathogen-induced alterations to the intestinal epithelial barrier and study microbial dissemination. By combining functional assays with cutting-edge imaging technologies, researchers can uncover critical mechanisms that drive microbial pathogenesis, revealing how pathogens breach intestinal defenses and spread systemically. This integrated approach deepens our understanding of infection dynamics and facilitates the development of new therapeutic strategies to combat gastrointestinal infections. Here, we describe a comprehensive approach to studying microbial pathogenesis by integrating in vitro and in vivo assays of intestinal cell adhesion, invasion, and translocation alongside advanced imaging techniques, such as confocal microscopy using immunostaining and fluorescence in situ hybridization (FISH) in tissue sections, to elucidate host–pathogen interactions and infection dynamics at the cellular level.

Key words Microbial pathogenesis, Intestinal epithelial cells, Cell adhesion, Invasion, Translocation, Host–pathogen interactions, Caco-2, Mouse and Gerbil models, Tight junction proteins, Confocal microscopy, Fluorescence in situ hybridization (FISH), Immunostaining, Z-stack imaging, Intestinal epithelial barrier, Infection dynamics

1 Introduction

The gastrointestinal (GI) tract is a primary interaction site between the host and many microbes, including commensal and pathogenic microorganisms [1]. The ability of specific pathogens to adhere to, invade, and translocate through intestinal epithelial cells is central

Inka Brockhausen (ed.), *Dynamics of Bacteria-Mucus Interactions*, Methods in Molecular Biology, vol. 2942,
https://doi.org/10.1007/978-1-0716-4627-4_1,

to their pathogenic mechanisms that are critical to the initiation and progression of infections. Understanding these mechanisms is vital to unraveling microbial pathogenesis and the host's defense strategies [1]. Both in vitro and in vivo models, alongside advanced imaging techniques such as confocal microscopy and fluorescence in situ hybridization (FISH), offer powerful tools to study these interactions in detail. Here, we outline how these approaches contribute to understanding microbial behavior and host–pathogen interactions, with a particular emphasis on intestinal cell adhesion, invasion, and translocation.

1. *Intestinal Cell Adhesion, Invasion, and Translocation:*

Adhesion to intestinal epithelial cells is the first critical step in many enteric infections [2]. Pathogens must successfully adhere to host cells to establish infection and resist clearance mechanisms, such as peristalsis, sloughing cells, and mucous exocytosis. The adhesion process involves the engagement of specific bacterial surface molecules—such as pili, adhesins, and outer membrane proteins interaction with receptors on the host cell surface [2]. This initial attachment secures the pathogen colonization and triggers signaling cascades that can alter the host cell environment, facilitating subsequent invasion [1, 3].

Invasion occurs when pathogens actively penetrate the intestinal epithelial barrier, gaining access to underlying tissues or circulating immune cells [1, 4]. This process is crucial for pathogens that cause systemic infections, such as *Listeria monocytogenes*, *Salmonella enterica* serovar Typhimurium, and *Shigella flexneri* [2]. Pathogens use a variety of mechanisms to invade host cells, often involving actin polymerization or the hijacking of host cell endocytosis pathways. For example, *Listeria* uses the *Listeria* Adhesion Protein (LAP)-Heat Shock Protein 60 (Hsp60) interaction and the internalins (InlA/E-cadherin and InlB/c-Met) to engage host cell receptors, respectively, facilitating its internalization [1, 4–7].

Once pathogens successfully invade epithelial cells, the next phase in pathogenesis is translocation across the intestinal barrier [2]. This can occur through transcellular pathways, where microbes traverse through epithelial cells, or paracellular pathways by breaching the tight junctions between cells [1]. Pathogen translocation is central to their ability to reach subepithelial compartments (lamina propria), where they can evade immune surveillance and disseminate to systemic sites, causing widespread infections. The mechanisms governing this process, such as the disruption of tight junction proteins (e.g., occludin and claudin), are critical areas of investigation, as they offer insights into how pathogens breach host barriers [1].

In vitro intestinal cell models, such as Caco-2 and HT-29 cell lines, allow researchers to study adhesion, invasion, and translocation in a controlled environment [2]. These epithelial cell models mimic the intestinal barrier and provide insights into pathogen–host interactions at the cellular level. For example, assays measuring bacterial adhesion and invasion can quantify the number of pathogens adhering to or internalized by epithelial cells. In contrast, translocation assays can assess bacterial movement across polarized monolayers. These experiments are often complemented by immunostaining techniques to visualize specific host and microbial factors involved in the interaction process.

2. *In Vivo Models:*

While in vitro studies are invaluable due to the controlled environment they provide, they do not fully recapitulate the complex interactions between host tissues, immune responses, and the microbiome in a living organism. In vivo models, particularly rodents (mice, gerbils, guinea pigs), are indispensable for studying microbial pathogenesis under physiological conditions [2, 8]. Mouse models offer several advantages, such as observing infection dynamics over time, monitoring systemic spread, and analyzing tissue-specific immune responses. These models are particularly useful for studying microbial translocation across the intestinal epithelium and the subsequent invasion of internal organs such as the liver, spleen, kidney, brain, etc.

Mucosal adhesion is a critical factor in the pathogenicity of many enteric microbes, influencing their ability to colonize, invade, and persist within the host. Understanding the dynamics of microbial adherence and subsequent invasion requires a meticulous approach to isolate and enumerate bacteria residing in various intestinal compartments, including the mucus layer, epithelial cells, and lamina propria [4, 9]. This process begins by preparing ileal tissue, where a section of ileum is carefully collected and processed to eliminate luminal contents and reveal the internal surface. Following this, the mucus fraction is isolated using a buffer solution containing N-acetylcysteine, which effectively dislodges the mucus layer for analysis. Concurrently, epithelial cells are separated from the tissue, allowing for the enumeration of intracellular and extracellular bacteria using gentamicin treatment to eliminate any remaining extracellular organisms. Finally, the lamina propria fraction is isolated and processed to comprehensively characterize the microbial landscape within the intestinal environment. Researchers can accurately quantify microbial populations by employing plate counting techniques on selective agar plates, providing valuable insights into the mechanisms of mucosal adhesion and the pathogenesis during intestinal infection. This systematic approach is a foundational tool for exploring the intricate interactions between microbes and the host's mucosal surfaces.

Using genetically modified mice, researchers can study the role of specific host proteins or immune pathways in microbial pathogenesis. For instance, mice deficient in crucial tight junction proteins (e.g., myosin light chain kinase [MLCK], claudins, or occludin) can provide insight into how pathogens exploit these structures to enhance translocation [4, 7]. Moreover, the use of in vivo infection models enables the study of immune evasion strategies employed by pathogens and their interaction with immune cells.

3. *Immunolocalization and Advanced Imaging Techniques:*

Visualization of host–pathogen interactions at the cellular level is crucial for understanding the detailed mechanisms of microbial pathogenesis. Confocal microscopy is a powerful imaging tool that allows researchers to capture high-resolution images of immunostained tissue sections. This technique is widely used to study host–pathogen interactions in vitro and in vivo models. For example, confocal imaging of immunostained tissues from infected mice can reveal the localization of pathogens within the intestinal epithelium or submucosa and their association with host cell markers such as tight junction proteins or cytoskeletal elements [4, 7].

Confocal microscopy is also instrumental in studying changes in the architecture of the intestinal epithelium during infection. By labeling host proteins such as occludin, claudin-1, or E-cadherin, researchers can visualize disruptions to tight junctions or adherens junctions, which are often targeted by pathogens during translocation [1, 4, 7]. Fluorescently labeled antibodies allow for multichannel imaging, enabling researchers the simultaneous visualization of multiple targets within a single tissue section.

4. *Fluorescence In Situ Hybridization (FISH):*

FISH further enhances our ability to study microbial pathogenesis by enabling the visualization of bacterial or viral RNA/DNA within host tissues. This technique uses labeled oligonucleotide probes that hybridize to specific sequences in the microbial genome, allowing researchers to visualize the spatial distribution of pathogens within host tissues. FISH is particularly useful for studying mixed infections or the localization of commensal versus pathogenic microbes within the intestinal niche [5].

By combining imaging techniques like confocal microscopy (immunostaining and FISH) with functional assays such as bacterial adhesion, invasion, and translocation studies, researchers can gain comprehensive insights into the dynamics of microbial infection. These complementary approaches enable the detailed characterization of the spatial relationships between pathogens and host cells, the structural integrity of the epithelial barrier, and the role of host immune responses. Moreover, the integration of advanced 3D imaging and z-stack reconstruction techniques allows for the generation of detailed, three-dimensional models of host–pathogen interactions. This approach provides a complete understanding of the spatial and temporal dynamics of microbial pathogenesis within the intestinal epithelium that can reveal critical processes such as pathogen-induced epithelial damage, immune cell infiltration, and microbial dissemination.

Studying microbial pathogenesis at the intestinal level requires a multifaceted approach, incorporating both in vitro and in vivo assays to investigate microbial adhesion, invasion, and translocation alongside advanced imaging techniques like confocal microscopy and FISH. Together, these methodologies provide powerful insights into the complex interactions between host and pathogen, offering a deeper understanding of the molecular and cellular mechanisms that drive infection and disease. These insights are crucial for developing novel therapeutic strategies to combat enteric infections and improve human health.

2 Materials

2.1 Mucosal Adhesion/Invasion/Translocation by Plate Counting (CFU) Using Cell Lines

1. Caco-2 cell line (HTB-37, ATCC).
2. 24-well tissue culture plates.
3. Dulbecco's Modified Eagle Medium (DMEM) high glucose (4.5 g/L) with L-glutamine and sodium pyruvate supplemented with 10% fetal bovine serum (FBS; D10F) (HyClone).
4. Sterile phosphate-buffered saline (0.1 M PBS), pH 7.4.
5. Dulbecco's Phosphate-Buffered Saline (DPBS).
6. Transwell filter inserts (4-μm pore size) (Corning-Costar).
7. Voltmeter (Millipore-Sigma) for transepithelial electrical resistance (TEER) measurement.
8. Overnight-grown bacterial cultures (specific pathogens of interest).
9. 1.5 mL Eppendorf tubes.
10. Tryptic Soy Agar (TSA) plates (for plating bacteria) (BD BioSciences).
11. Luria Bertani broth (BD BioSciences).
12. Gentamicin sulfate (stock solution at 50 mg/mL; antibiotic for killing extracellular bacteria) (Sigma-Aldrich).
13. Triton X-100 (for lysing cells) (Sigma-Aldrich).

2.2 Mucosal Adhesion/Adhesion/Invasion/Translocation In Vivo by Plate Counting

1. 6 mM N-acetylcysteine solution (for mucus extraction) (Vivion, Inc).
2. RPMI 1640 medium (for epithelial cell processing) (HyClone).
3. 5 mM EDTA (for epithelial cell detachment).

4. 1 mM dithiothreitol (DTT) (for epithelial cell processing) (Sigma-Aldrich).
5. Type IV collagenase (for lamina propria digestion) (Worthington BioChem).
6. Gentamicin sulfate (25 mg/mL; antibiotic for killing extracellular bacteria).
7. Centrifuge (for pelleting cells and bacteria).
8. Cell culture incubator (37 °C, 5% CO_2).
9. Pipettes and sterile tips.
10. Vortex mixer (for resuspending pellets).
11. Plate counter or colony counter (for counting colonies).
12. Sterile tubes (for resuspending samples).
13. Sterile disposable pipettes.
14. Pipette tips.
14. 0.1 M sterile phosphate-buffered saline (pH 7.4).
15. Modified Oxford (MOX) or selective agar plates (Neogen).
16. FBS (fetal bovine serum) (HyClone).
17. Collagenase Type IV (1 mg/mL; Worthington).
18. DNase I (40 μg/mL; Worthington).
19. Centrifuge capable of low- and high-speed centrifugation.
20. Vortex.
21. Scalpel.
22. Shaker incubator.

2.3 Immunostaining and Confocal Imaging to Localize Pathogen in Tissue Sections

1. Rodent ileal and colonic tissue sections.
2. 10% formalin (Sigma-Aldrich).
3. Paraffin wax (Sigma-Aldrich).
4. 10 mM sodium citrate buffer (pH 6.0).
5. 0.01 M Tris-EDTA buffer (pH 9.0)
6. PBS (phosphate-buffered saline) (Sigma-Aldrich).
7. Triton X-100 (Sigma-Aldrich).
8. Normal goat serum (Cell Signaling).
9. Primary antibodies (e.g., MLCK, P-MLC, claudin-1, occludin, E-cadherin) (Thermo Fisher; Cell Signaling; Santa Cruz Biotech).
10. Alexa Fluor-488, Alexa Fluor-555, Alexa Fluor-647 secondary antibodies (Cell Signaling).
11. DAPI (4′,6-diamidino-2-phenylindole) (Thermo Fisher Scientific).

12. ProLong antifade reagent (Invitrogen).
13. Coverslips (1.5).
14. BSA (bovine serum albumin, Sigma-Aldrich).
15. Nikon A1R confocal microscope or Zeiss LSM 800 confocal microscope.

2.4 FISH and Confocal Imaging to Localize Commensals/Pathogen in Tissue Sections

1. Proteinase K (20 μL/mL in Tris-buffered saline with Tween 20).
2. TBST (Tris-buffered saline with Tween 20).
3. 2× saline sodium citrate (SSC) buffer (0.3 M NaCl and 0.03 M sodium citrate; anhydrous).
4. Graded ethanol (70%, 95%, 100%).
5. Alexa-488-conjugated Lcas467 *Lacticaseibacillus casei* DNA probe (sequence: 5′/5Alex488N/CCGTCACGCCGACAA-CAG-3′).
6. SSC buffers (2× SSC, 0.1× SSC).
7. DAPI (4′,6-diamidino-2-phenylindole; 1 μg/mL).
8. ProLong Gold antifade mounting medium.
9. Nikon A1R confocal microscope.
10. IQ Kinetic slide stainer (BioCare Medical).
11. Slides with paraffin-embedded tissue.
12. Coverslips (1.5).

3 Methods

3.1 Mucosal Adhesion Analysis of Bacteria Using Cell Lines by Plate Counting

3.1.1 Growing Intestinal (Cell Lines) to a Monolayer on a Tissue Culture Plate

1. Intestinal cell lines such as Caco-2 are cultured using DMEM (Dulbecco's Modified Eagle Medium)—high glucose (4.5 g/L) with L-glutamine and sodium pyruvate containing 10% fetal bovine serum (FBS) (D10F) (*see* **Note 1**).
2. Seed Caco-2 cells into a 24-well tissue culture plate (Corning-Coster). A typical seeding density is 0.05×10^6 cells/24 well. Add 0.5 mL of D10F medium to each well.
3. Incubate the cells at 37 °C in a humidified 5% CO_2 incubator maintaining a 5% CO_2 level until the cells reach a confluent monolayer and are polarized. This is generally achieved 14–21 days post-seeding, and a typical cell number is ~5×10^5 cells/well (*see* **Note 2**).

3.1.2 Preparation of Bacterial Cultures

1. On the day of the experiment, obtain 1 mL of overnight-grown bacterial cultures in stationary phase (12–16 h) in a cuvette.
2. Blank the spectrophotometer with a sample of the sterile medium (e.g., LB, TSB) to remove background noise.

3. Place a small aliquot (1 mL) of the bacterial culture in a cuvette. Measure the OD at 600 nm (OD600). Record the OD reading and correlate OD600 to CFU/mL in the previously created standard curve (*see* **Note 3**).
4. Obtain 1 mL of overnight-grown bacterial cultures (12–16 h) in a 1.5 mL Eppendorf tube.
5. Centrifuge the bacterial cultures at 8000×*g* for 2–3 min to pellet the cells.
6. Carefully discard the supernatant without disturbing the pellet.
7. Wash the bacterial pellet by adding 1 mL of sterile PBS and resuspending the cells. Repeat this step three times.
8. Resuspend the bacterial pellet in 1 mL of D10F medium.
9. The expected concentration for Gram-positive bacterial pathogens such as *Listeria monocytogenes* is approximately 5×10^8 colony forming units (CFU)/mL (5×10^7 CFU/100 μL).

3.1.3 Adhesion Assay

1. Wash the Caco-2 monolayers once by adding 1 mL of DPBS to each well.
2. Gently aspirate and discard the DPBS.
3. Add 0.4 mL of fresh D10F medium to each well.
4. Add 100 μL of the washed bacterial culture to the corresponding well containing the Caco-2 monolayer.
5. This step results in a multiplicity of infection (MOI) of ~100 (~5×10^7 CFU/5×10^5 mammalian cells) (*see* **Note 4**).
6. Incubate the infected Caco-2 monolayers for 1 h at 37 °C in a 5% CO_2 incubator (*see* **Note 5**).
7. During incubation, perform tenfold serial dilutions of the original overnight (12–16 h) grown bacterial inoculum and plate the diluted samples on TSA agar plates to determine CFU/mL.
8. After incubation, wash the monolayers three times with 1 mL of DPBS to remove the nonadherent bacteria.
9. Remove the monolayers and the adherent bacteria by adding 1 mL of 0.1% Triton X-100 for 5–10 min. Prepare prior by adding 50 μL of Triton X-100 in 50 mL of D10F media. Mix the lysate using a vortex mixer.
10. Perform tenfold serial dilutions of the lysed samples in sterile PBS.
11. Plate the diluted samples on TSA agar plates.
12. Incubate the TSA plates at the appropriate temperature for bacterial growth (typically 37 °C for 24–48 h).

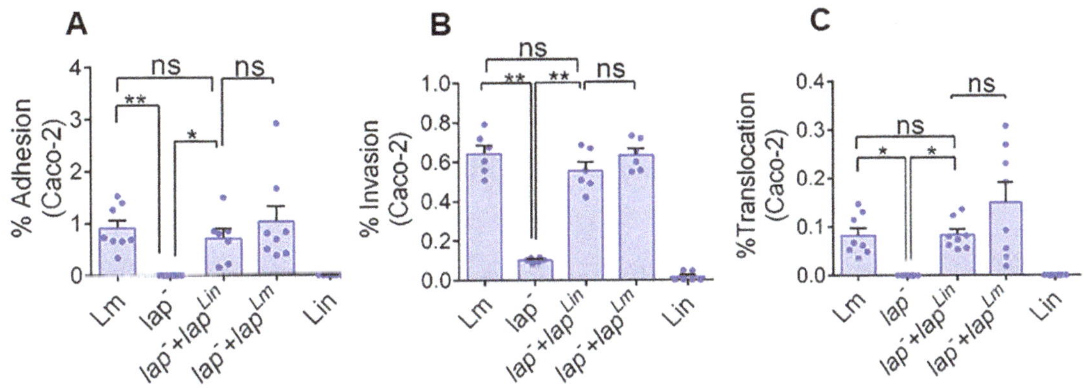

Fig. 1 (**a**–**c**) Restoration of adhesion (MOI 10, 1 h) (**a**) invasion (MOI 10, 2 h) (**b**), and translocation (MOI 10, 2 h) of the *lap*$^{-}$ + *lap*^{Lin} strain in Caco-2 cells. The heterologous complementation (*lap*$^{-}$ + lap^{Lin}) restored the ability of the lap^{-} strain to adhere and translocate across the human enterocyte-like Caco-2 cell monolayers to levels similar to those of the wild-type (WT) Lm strain (F4244, serovar 4b, clinical strain) or the homologous complemented lap deficient *Lm* strain (*lap*$^{-}$ + lap^{Lm}). *L. innocua*, a nonpathogenic strain, is used as a negative control

13. Count the colonies after incubation to determine the number of viable bacteria from each sample and inoculum.
14. Calculate adhesion (%) = (CFU on the surface after adhesion per well/initial CFU added per well/) × 100 (*see* **Note 6** and Fig. 1a).

3.1.4 Invasion Assay

1. On the day of the experiment, wash the Caco-2 monolayers in the well (cultured as described above) once by adding 1 mL of DPBS to each well (*see* **Note 1**).
2. Gently aspirate and discard the DPBS.
3. Add 0.4 mL of fresh D10F medium to each well.
4. Add 100 μL of the washed bacterial overnight (12–16 h) culture to the corresponding well containing the Caco-2-monolayer (*see* **Note 2**).
5. This step results in a multiplicity of infection (MOI) of ~100 (~5 × 10^7 bacterial cells/~5 × 10^5 mammalian cells) (*see* **Note 4**).
6. Incubate the infected Caco-2 monolayers for 1 h at 37 °C in a 5% CO_2 incubator (*see* **Note 5**).
7. During incubation, perform tenfold serial dilutions of the original bacterial O/N inoculum and plate the diluted samples on TSA agar plates to determine CFU/mL.
8. Add 1 mL of D10F medium containing 50 μg/mL gentamicin sulfate to each well to kill any remaining extracellular bacteria.
9. Incubate the plate for 30 min at 37 °C in a 5% CO_2 incubator.

10. Wash the monolayers thrice with 1 mL of DPBS to remove nonadherent bacteria.
11. Discard the DPBS after each wash.
12. Add 1 mL of 0.1% Triton X-100 to each well to lyse the Caco-2 cells and release internalized bacteria.
13. Perform tenfold serial dilutions of the lysed samples in sterile PBS.
14. Plate the diluted samples on TSA agar plates.
15. Incubate the TSA plates at the appropriate temperature for bacterial growth (typically 37 °C for 24–48 h).
16. Count the colonies after incubation to determine the number of viable bacteria from each sample and inoculum.
17. Calculate invasion (%) = (CFU after gentamicin protection per well/initial CFU added per well/) × 100. (*see* **Notes 6** and 7 and Fig. 1b).

3.1.5 Translocation Assay

1. Caco-2 cells are grown to confluence on transwell filter inserts with a 4-μm pore size.

   Seeding density for Caco-2 on transwells: (*see* **Notes 1** and **2**).
2. Transwell insert diameter:
   - 6.5 mm (24-well plate) → 0.33 cm^2 surface area.
   - 12 mm (12-well plate) → 1.12 cm^2 surface area.
   - 24 mm (6-well plate) → 4.67 cm^2 surface area.
3. Typical seeding densities:
   - 24-well plate (0.33 cm^2): 50,000–100,000 cells per insert.
   - 12-well plate (1.12 cm^2): 200,000–250,000 cells per insert.
   - 6-well plate (4.67 cm^2): 1,000,000–1200,000 cells per insert.
4. Prepare a single-cell suspension of Caco-2 cells after trypsinization.
5. Count the cells using a hemocytometer or an automated cell counter.
6. Dilute the cell suspension to achieve the desired cell concentration for the appropriate well size.
7. Add the required volume of cell suspension to the apical chamber of each transwell insert, ensuring even distribution of cells.
8. Typically, the medium is added to the apical (insert) and basolateral (well) chambers for proper culture conditions. For a 12-well plate, 0.5 mL of medium is added to the apical side and 1.5 mL to the basolateral side.

9. Transepithelial electrical resistance (TEER) is measured to assess the integrity of the polarized monolayer.
10. Use a voltmeter (Millipore) to measure TEER.
11. Criteria: Only monolayers with a TEER value of around 200 Ω/cm^2 (range: 190–209 Ω/cm^2) are used for further experiments. This indicates a well-formed tight junction with an intact epithelial barrier.
12. Bacteria are added to the apical well of the transwell system at a multiplicity of infection (MOI) of approximately 10–100, as described in adhesion and invasion protocols (*see* **Note 4**).
13. The setup is incubated at 37 °C in a 5% CO_2 atmosphere for 2 h (*see* **Note 5**).
14. Enumerate the translocated bacteria by serially diluting media collected from the basolateral chamber and plating on TSA agar plates.
15. Calculatetranslocation (%) = (initial CFU in apical chamber/ CFU in basolateral chamber) × 100 (*see* **Note 6** and Fig. 1c).

3.2 Mucosal Adhesion In Vivo by Plate Counting Method

3.2.1 Preparation of Ileal Tissue

1. Obtain a 10 cm section of ileum tissue from uninfected or infected rodents (mice/rats/gerbil/guinea pigs) (*see* **Note 8**).
2. Flush the ileum section with sterile PBS to remove luminal content.
3. Remove any visible Peyer's patches from the ileum using a scalpel (*see* **Note 9**).
4. Cut the ileum longitudinally to expose the internal surface.

3.2.2 Isolation of Mucus Fraction

1. Prepare a solution of 6 mM N-acetylcysteine (6 mM) in sterile PBS.
2. Place the ileum tissue into a sterile tube containing 3 mL of the N-acetylcysteine solution.
3. Incubate the tissue in the solution for 2 min (*see* **Note 10**).
4. After incubation, shake the tube vigorously to dislodge the mucus layer.
5. Transfer the tissue to a fresh tube with 3 mL of N-acetylcysteine solution and repeat the wash two more times (total of three washes).
6. Pool the washes (total volume = 9 mL) into a single tube (*see* **Note 11**).
7. Centrifuge the pooled mucus wash at 12,000×*g* for 20 min.
8. Resuspend the resulting pellet in 0.5 mL of PBS.
9. Vortex the suspension briefly to ensure even mixing.

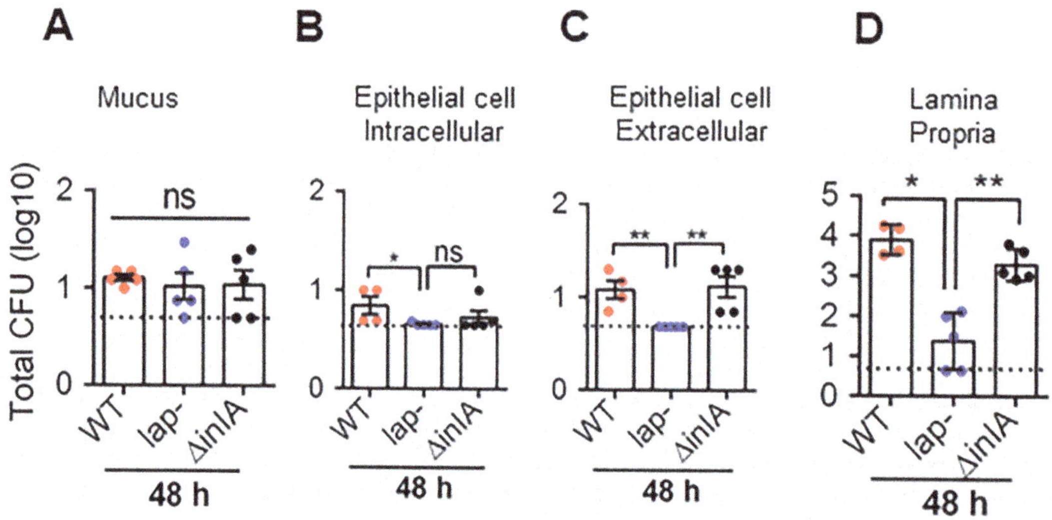

Fig. 2 (**a–d**) *Listeria* counts (total CFU), ileal mucus (**a**), epithelial cell intracellular (**b**), epithelial cell-extracellular (**c**), and lamina propria (**d**; intracellular and extracellular) of A/J mice ($n = 4–6$) at 48 h postinfection from three independent experiments. Dashed horizontal lines indicate the detection limit

10. Perform serial dilutions of the suspension and plate on selective agar plates (MOX for counting *L. monocytogenes*).
11. Incubate the plates and count colonies to enumerate bacteria in the mucus fraction (total, Fig. 2a).

3.2.3 Isolation of Epithelial Cell Fraction

1. Cut the ileal tissue into small pieces (approximately 1 cm each) using a scalpel.
2. Prepare a solution of RPMI medium supplemented with 5 mM EDTA and 1 mM DTT (*see* **Note 12**).
3. Place the tissue pieces into a sterile tube containing 5 mL of the RPMI-EDTA-DTT solution.
4. Incubate the tube at 37 °C with shaking for 10 min.
5. Transfer the tissue pieces to a fresh tube with 5 mL of RPMI-EDTA-DTT solution and repeat the incubation two more times (total of three washes).
6. Pool the washes (total volume = 15 mL) into a single tube (*see* **Note 13**).
7. Centrifuge the pooled epithelial cell wash at 1200×*g* for 10 min to pellet the cells.
8. Separate the supernatant from the pellet for further processing.

3.2.4 Enumeration of Intracellular Bacteria in Epithelial Cell Fraction

1. Resuspend the cell pellet from the epithelial fraction in 5 mL RPMI medium containing 25 mg/mL of gentamicin sulfate.
2. Incubate the suspension at 37 °C with 5% CO_2 for 30 min (cell culture incubator) to kill any extracellular bacteria (*see* **Note 14**).

3. Centrifuge the suspension at 1200×*g* for 10 min to pellet the cells.
4. Wash the pellet twice with sterile PBS.
5. Resuspend the washed pellet in 0.5 mL of PBS.
6. Vortex briefly to lyse the cells (*see* **Note 15**).
7. Perform serial dilutions and plate on selective agar plates. MOX agar plates are used for *Listeria monocytogenes.*
8. Incubate the plates and count colonies to enumerate intracellular bacteria in the epithelial cell fraction (Fig. 2b).

3.2.5 Enumeration of Extracellular Bacteria in Epithelial Cell Fraction

1. Centrifuge the supernatant from the epithelial cell fraction washes at 12,000×*g* for 20 min.
2. Resuspend the resulting pellet in 0.5 mL of PBS.
3. Vortex briefly to ensure even mixing.
4. Perform serial dilutions and plate on selective agar plates (MOX for *Listeria monocytogenes*).
5. Incubate the plates and count colonies to enumerate extracellular bacteria in the epithelial cell fraction (Fig. 2c).

3.2.6 Isolation of Lamina Propria Fraction

1. Wash the tissue pieces (after removal of epithelial cells) twice with 25 mL of sterile PBS to remove EDTA and DTT.
2. Prepare a digestion solution consisting of RPMI medium supplemented with 5% FBS, 1 mg/mL Type IV collagenase, and 40 μg/mL DNase I (*see* **Note 16**).
3. Place the tissue pieces into a sterile tube containing 4 mL of the digestion solution.
4. Incubate at 37 °C with shaking for 40 min.
5. Repeat the digestion step in a fresh tube with the digestion solution until the tissue pieces are completely dissolved.
6. Pool the digestion solutions into a single tube.
7. Centrifuge the pooled digestion solution at 1200×*g* for 10 min to pellet the cells.
8. Separate the supernatant from the pellet for further processing.

3.2.7 Enumeration of Intracellular Bacteria in Lamina Propria Fraction

1. Resuspend the cell pellet from the lamina propria fraction in 5 mL of RPMI medium containing 25 mg/mL gentamicin (*see* **Note 14**).
2. Incubate the suspension at 37 °C with 5% CO_2 for 30 min (cell culture incubator) to kill any extracellular bacteria.
3. Centrifuge the suspension at 1200×*g* for 10 min to pellet the cells.
4. Wash the pellet twice with sterile PBS.

5. Resuspend the washed pellet in 0.5 mL of PBS.
6. Vortex briefly to lyse the cells.
7. Perform serial dilutions and plate on selective agar plates (MOX agar for *L. monocytogenes*).
8. Incubate the plates and count colonies to enumerate intracellular bacteria in the lamina propria fraction (Fig. 2d).

3.2.8 Enumeration of Extracellular Bacteria in Lamina Propria Fraction

1. Centrifuge the supernatant from the lamina propria digestion at 12,000×*g* for 20 min.
2. Resuspend the resulting pellet in 0.5 mL of PBS.
3. Vortex briefly to ensure even mixing.
4. Perform serial dilutions and plate on selective agar plates (MOX agar plates for *L. monocytogenes*).
5. Incubate the plates and count colonies to enumerate extracellular bacteria in the lamina propria fraction (Fig. 2d).

3.3 Immunostaining and Confocal Imaging to Localize Pathogen in Tissue Sections

3.3.1 Tissue Preparation

1. Collect ileal and colonic tissue sections from uninfected or infected rodents (*see* **Note 8**).
2. Fix the tissues in 10% neutral-buffered formalin at room temperature for 24–48 h.

3.3.2 Embedding

1. After fixation, dehydrate the tissues in increasing concentrations of ethanol (70%, 90%, and 100%, each for 3–5 min), followed by xylene treatment.
2. Embed the tissues in paraffin wax (*see* **Note 17** for use of a central histopathology facility).

3.3.3 Sectioning

1. Section the paraffin-embedded tissue into 5 μm thick slices using a microtome (*see* **Note 17** for the use of a central histopathology facility).
2. Mount the sections on glass slides.

3.3.4 Deparaffinization and Rehydration

1. Place the tissue slides in xylene for 2 × 5 min.
2. Immerse the slides in 100% ethanol for 2 × 3 min.
3. Gradually rehydrate by immersing the slides in 95%, 70%, and 50% ethanol for 3 min each.
4. Wash the slides in distilled water.

3.3.5 Antigen Retrieval

1. Heat the tissue sections by immersing them in boiling sodium citrate buffer (10 mM, pH 6.0) or Tris-EDTA buffer (0.01 M, pH 9.0) for 10 min (Effective for more resistant antigens that require more aggressive unmasking).

2. Allow the slides to cool at room temperature for 20 min.
3. Rinse the slides in PBS for 5 min.

3.3.6 Permeabilization, Blocking, Antibody Incubation, and Mounting

1. Incubate the tissue sections in PBS containing 0.3% Triton X-100 for 10 min at room temperature to permeabilize the tissue. Use pap pen to create hydrophobic barrier (*see* **Note 18**).
2. Block the sections in PBS containing 5% normal goat serum for 1 h at room temperature to prevent nonspecific binding.
3. Primary antibody incubation: Prepare a single or cocktail of primary antibodies at the appropriate dilutions: *Lm* (1:100, our lab or commercially available), MLCK (1:100; RRID: AB_477243), P-MLC (1:200; RRID: AB_2250969), claudin-1 (1:200; RRID: AB_2533323), occludin (1: 150; RRID: AB_2533101), ZO-1 (1:100; RRID: AB_628459), and E-cadherin (1:200; RRID: AB_2291471) (*see* **Note 18**).
4. Incubate the tissue sections with the primary antibody cocktail overnight at 4 °C.
5. Wash the slides three times with PBS (3 cycles, 5 min per wash).
6. Incubate the tissue sections with a single or a cocktail of secondary antibodies conjugated to a fluorophore (e.g., Alexa Fluor-488 and Alexa Fluor-555).
7. Incubate for 2 h at room temperature in the dark (*see* **Note 19**).
8. For nuclear staining, wash the slides three times with PBS, and incubate the tissue sections with DAPI (1 μg/mL in PBS) for 5 min to stain the nuclei.
9. Wash the slides with PBS and mount them using ProLong antifade reagent using a 1.5 coverslip.
10. Allow the mounting medium to cure overnight at room temperature in the dark.

3.3.7 Confocal Microscopy

Acquire images using a Nikon A1R or Zeiss LSM 800 confocal microscope equipped with appropriate lasers for excitation (405 nm, 488 nm, 561 nm, and 640 nm).

1. Image acquisition: Use a 60× or 63× oil immersion objective (1.40 NA).
2. Perform z-stack scanning with 0.25 μm intervals through the 5 μm-thick sections.
3. Capture X-Z and Y-Z cross-sections using orthogonal reconstruction from the z-stack.
4. Perform three-dimensional reconstructions using the Nikon Elements or Zen 5 software.

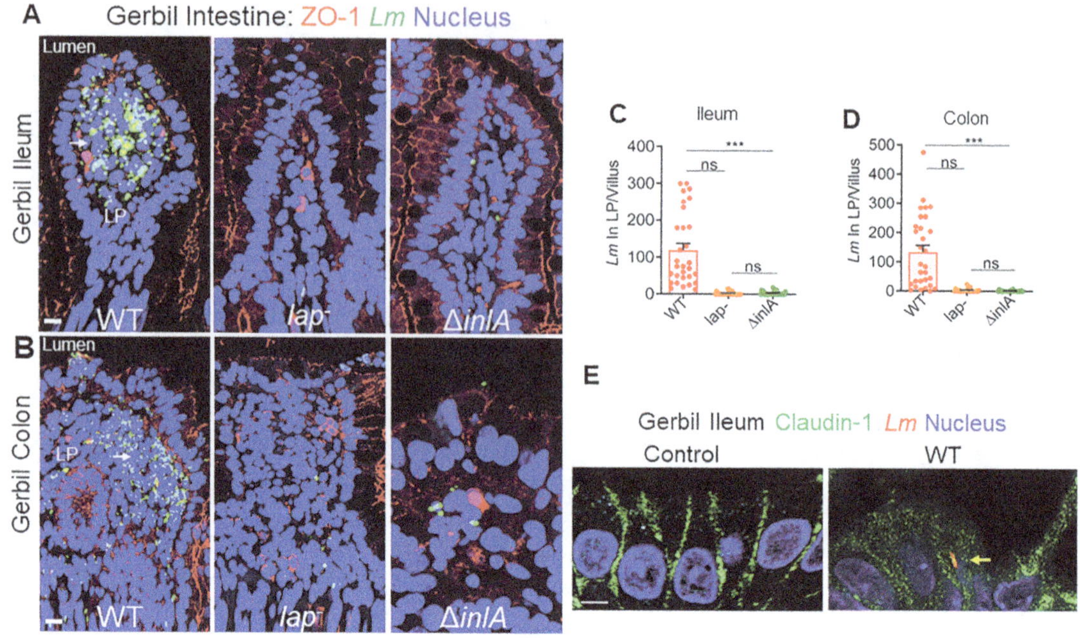

Fig. 3 (**a** and **b**) Representative confocal immunofluorescence microscopic images of ileal (**a**) or colonic (**b**) tissue sections immunostained for ZO-1 (red), *Listeria* (green), and 4′,6-diamidino-2-phenylindole (DAPI, blue; nucleus) from WT, *lap*⁻, *ΔinlA*,−*L. monocytogenes* bacteria-challenged gerbils at 48 hpi, bars, 10 μm. Increased *L. monocytogenes* (green) was detected in the lamina propria of ileal (A) or colonic (B) tissue in WT-challenged gerbils (arrows). (**c** and **d**) Graph representing quantitative measurements of *L. monocytogenes* counts in the lamina propria from ileal (I) or colonic (J) villi images ($n = 30$ villi) from three gerbils for each treatment. (**e**) Representative confocal immunofluorescence picto-micrographs of the gerbil ileal tissue sections immunostained for claudin-1 (green), *Listeria* (red; yellow arrow), and DAPI (blue; nucleus) from uninfected control or WT challenged gerbil at 48 hpi. Bars, 5 μm. Lm (yellow arrow) cells are seen in lamina propria in WT with mislocalization of claudin-1 (intracellular puncta, endocytosis with WT). Images are representative of five different fields from three gerbils. *LP* lamina propria

3.3.8 Post-acquisition Processing

Perform post-acquisition image processing using Adobe Photoshop or other image processing software for adjustment and analysis (*see* **Note 20** and Figs. 3a–e).

3.4 Immunostaining and Confocal Imaging to Localize Commensals/Pathogen Using Fluorescent In Situ Hybridization (FISH)

3.4.1 Deparaffinization

1. Place the paraffin-embedded tissue slides in a slide rack.
2. Deparaffinize the slides using the following steps.
3. Immerse the slides in xylene for two to three changes (5 min each) to remove the paraffin.
4. Hydrate the slides by sequential immersion in graded ethanol (100%, 95%, 70%) for 3–5 min each.
5. Rinse slides in distilled water to remove ethanol.

3.4.2 Enzyme Digestion with Proteinase K

1. Prepare the Proteinase K solution at 20 μL/mL in TBST (Tris-buffered saline with Tween 20).
2. Load the slides into the BioCare Medical IQ Kinetic slide stainer.
3. Incubate the slides with the prepared Proteinase K solution at 37 °C for 10 min. This step will digest proteins and enhance probe accessibility.
4. After digestion, wash the slides with 2× SSC buffer containing 0.1% Tween 20 to remove excess enzyme and buffer components.

3.4.3 Dehydration

1. Dehydrate the tissue sections by sequential immersion in graded ethanol:

   70% ethanol for 3 min.

   95% ethanol for 3 min.

   100% ethanol for 3 min.
2. Allow the slides to air-dry completely.

3.4.4 Probe Hybridization

1. Dilute the Alexa-488-conjugated *L. casei* DNA probe (Lcas467) to a concentration of 0.11 ng/μL (1:60 dilution) (*see* **Notes 21** and **22**).
2. Apply the probe to the air-dried slides.
3. Denature the oligomer and the DNA on the tissue sections by incubating the slides at 78 °C for 5 min.
4. Hybridize the probe by incubating the slides at 45 °C for 16 h in a humidified chamber. This allows the probe to bind to the complementary target sequences.

3.4.5 Post-hybridization Washes

1. Wash the slides with 2× SSC buffer to remove the excess unbound probe.
2. Perform a stringency wash with 0.1× SSC at 60 °C to eliminate any nonspecific binding of the probe.
3. Wash the slides again with 2× SSC to remove the residual SSC buffer and ensure the sections are clean.
4. Counterstaining with DAPI: Incubate the slides with DAPI (1 μg/mL) for 5 min to stain the nuclei. Rinse the slides briefly with distilled water to remove excess DAPI.

3.4.6 Mounting

1. Mount the slides with ProLong Gold antifade reagent to prevent photobleaching.
2. Place a coverslip (1.5) gently over the tissue sections.

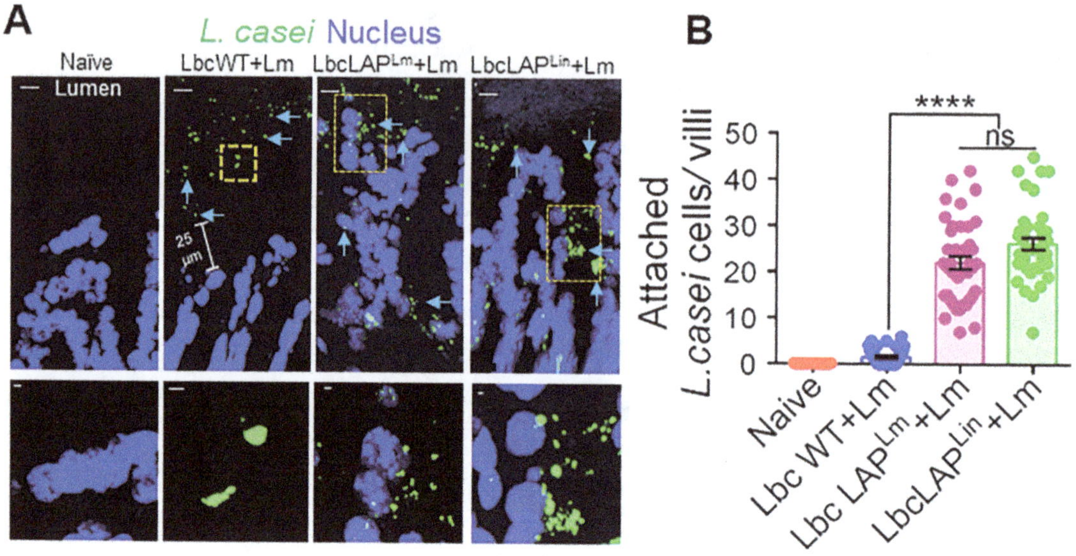

Fig. 4 (**a**) Micrographs of colonic villi of LbcWT or BLP-treated mice (10 days) at 48 hpi after fluorescent in situ hybridization (FISH, c) using *L. casei(Lbc)*-specific 16 s rDNA probe (green) and counterstained to visualize the nucleus (DAPI). In LbcWT-treated mice, bacteria were seen in the mucus layer (arrows) in the lumen, while bioengineered Lactobacillus probiotic (BLP) expressing LAP-treated mice bacterial clusters (arrows) are primarily on the surface of epithelial cells. Bars, 10 μm. The enlargements of the boxed areas in each image are shown in the bottom panel. Bars, 1 μm. (**b**) Quantitative measurements (mean ± SEM) of attached *L. casei* cells to the epithelial cells from FISH images. Each point represents an individual villus, $n = 40$ villi from four mice for each treatment

3.4.7 Image Acquisition

1. Visualize the slides using a Nikon A1R confocal microscope.
2. Acquire images under appropriate fluorescence channels to detect Alexa-488-labeled probes and DAPI-stained nuclei (Fig. 4a).
3. Analyze the captured images for probe hybridization efficiency and localization (*see* **Note 20** and Figs. 4a, b).

4 Notes

1. Cell line selection: Choose a cell line that closely mimics the target host tissue, e.g., Caco-2 or HT-29 cells for intestinal epithelium or primary intestinal cells for more physiological relevance.
2. Confluency and differentiation: For studies requiring differentiation (e.g., barrier assays), Caco-2 cells typically take 2–3 weeks to form a tight monolayer after seeding, during which time the medium should be replaced every 2–3 days.
3. Create a standard curve or refer to an already established correlation for the specific bacteria and growth conditions you work with.

4. Multiplicity of infection (MOI): Determine the appropriate MOI to avoid overwhelming the host cells. Depending on the pathogen and assay design, the MOI is typically between 10 and 100 bacteria per host cell.
5. Assay duration: Optimize the incubation time for pathogen–host interactions to prevent premature detachment or overgrowth of pathogens.
6. Washing steps: After incubation, wash the cells thoroughly to remove nonadherent pathogens. The washing method and buffer must be gentle to avoid disrupting the adherent bacteria.
7. Controls, replicates, and strain variation: Include proper controls such as uninfected cells, heat-killed pathogens, or closely related nonpathogen (noninvasive strains) to validate assay specificity. Always perform assays in biological replicates (at least three to four independent) to account for variability in cell culture and bacterial preparations. Invasion ability can differ widely between pathogen strains (clinical/environmental/food), so ensure that the tested strain is relevant to the disease context.
8. Source selection/ethical compliance/timing postinfection: Ensure the selection of appropriate animal models (mice, rats, gerbils, guinea pigs) based on the study's objectives and pathogen relevance. Adhere to institutional and national ethical guidelines for animal handling and tissue collection. If using infected rodents, standardize the time postinfection when tissues are harvested to capture consistent stages of infection.
9. Carefully excise Peyer's patches with a scalpel to avoid damaging the surrounding mucosal tissue, which could affect bacterial distribution.
10. N-Acetylcysteine concentration and timing: Ensure precise concentration and incubation time with N-acetylcysteine to effectively dislodge the mucus layer without affecting the epithelial cells beneath. Too long or harsh incubation may detach epithelial cells, skewing results.
11. Carefully pool and centrifuge the washes to ensure an adequate yield of mucus without debris. Ensure consistent centrifugation speed and time to achieve proper pellet formation.
12. EDTA and DTT are crucial for loosening epithelial cells. However, too high concentrations or prolonged incubation may damage cells or alter their integrity, affecting the assay.
13. Washing steps: Consistency in washing and pooling steps is critical for reproducibility. Ensure that all washes are collected carefully without losing any cell fractions.

14. Gentamicin protection assay: Gentamicin must effectively kill extracellular bacteria without entering or affecting intracellular bacteria. Proper incubation times and gentamicin concentrations are essential for accuracy.
15. Cell lysis: Ensure gentle but effective lysis of epithelial cells to release intracellular bacteria without damaging bacterial integrity. Use brief vortexing to avoid excessive cell damage.
16. The use of collagenase and DNase is crucial for breaking down the tissue matrix. Ensure enzyme concentrations and incubation times are optimized to release cells without causing excessive degradation or loss of viable bacteria.
17. Paraffin embedding and microtomy: Once tissues are fixed, they are dehydrated and embedded in paraffin to create tissue blocks. A central pathology facility uses automated tissue processors to move samples through ethanol, xylene, and paraffin stages in a controlled manner. Microtomy for paraffin sections: The facility uses microtomes to slice thin sections (usually 4–6 μm) of paraffin-embedded tissues. These sections are placed on slides for staining and microscopic examination. The precision of sectioning is critical for maintaining tissue architecture and ensuring accurate analysis.
18. The pap pen is a crucial tool used in immunohistochemistry (IHC) and immunofluorescence (IF) procedures to create a hydrophobic barrier around tissue sections on a slide. This barrier helps contain the small volumes of reagents (e.g., antibodies, blocking solutions, or detection reagents) applied during staining, preventing them from spreading over the slide and mixing with other sections. By focusing the reagents on the tissue area of interest, the pap pen ensures efficient use of antibodies, reduces reagent waste, and improves staining uniformity.
19. Primary and secondary antibodies: If multiplexing, choose antibodies targeting different antigens, raised in different host species (e.g., rabbit, mouse) to prevent cross-reactivity.
20. Secondary antibodies: For visualization, use species-specific secondary antibodies (e.g., anti-mouse, anti-rabbit) conjugated to different fluorophores (for example, Alexa-flour 488 and 555).
21. If quantifying bacteria in specific areas (e.g., within tissue sections), draw region of interest (ROIs) around those areas and count bacteria by using one of the following methods:
    A. Manual counting: Visually inspect the images and manually count each bacterium, making sure to account for potential overlap or clumping.

B. Image analysis software: Utilize software like ImageJ or Fiji to assist in counting. Use the "Analyze Particles" feature to count objects above the defined threshold.

22. Designing FISH probes for 16S rRNA involves selecting specific target regions within the gene, typically focusing on hypervariable areas that differentiate bacterial species. Researchers should retrieve and analyze 16S rRNA sequences from databases like NCBI or Silva, using multiple sequence alignments to pinpoint conserved and variable sequences. Probes should be 20–50 nucleotides long, with a GC content of 40–60% to ensure optimal binding and specificity for the target organism. After selecting a fluorescent dye for labeling, it's essential to validate the probe's hybridization efficiency and specificity through in silico tools like BLAST. Finally, experimental conditions must be optimized for successful visualization using fluorescence microscopy.

Acknowledgements

The research was supported by funds from the National Institute of Food and Agriculture of USDA (Hatch grant # 1016249; A.K.B.), KY-INBRE (part of award # P20GM103436-22 from the National Institute of General Medical Sciences; R.D.), the start-up funds at ODU (R.D.), and USDA-NIFA (Award # 2023-67017-40051; R.D.).

References

1. Drolia R, Bhunia AK (2019) Crossing the intestinal barrier via *Listeria* adhesion protein and internalin a. Trends Microbiol 27:408–425
2. Bhunia AK (2018) Foodborne microbial pathogens: mechanisms and pathogenesis. Springer Nature, New York
3. Ireton K, Van Ngo H, Bhalla M (2018) Interaction of microbial pathogens with host exocytic pathways. Cell Microbiol 20:e12861
4. Drolia R, Tenguria S, Durkes AC, Turner JR, Bhunia AK (2018) *Listeria* adhesion protein induces intestinal epithelial barrier dysfunction for bacterial translocation. Cell Host Microbe 23:470–484
5. Drolia R, Amalaradjou MAR, Ryan V et al (2020) Receptor-targeted engineered probiotics mitigate lethal *Listeria* infection. Nat Commun 11:6344
6. Liu D, Bai X, Helmick HDB, Samaddar M et al (2023) Cell-surface anchoring of *Listeria* adhesion protein on *L. monocytogenes* is fastened by internalin B for pathogenesis. Cell Rep 42: 112515
7. Drolia R, Tenguria S, Bryant DB et al (2024) *Listeria* adhesion protein orchestrates caveolae-mediated apical junctional remodeling of epithelial barrier for *Listeria monocytogenes* translocation. MBio 15:e02821–e02823
8. Sousa M, Magalhães R, Borges Ferreira V, Teixeira P (2024) Current methodologies available to evaluate the virulence potential among *Listeria monocytogenes* clonal complexes. Front Microbiol 15:1425437
9. Bou Ghanem EN, Jones GS, Myers-Morales T et al (2012) InlA promotes dissemination of *Listeria monocytogenes* to the mesenteric lymph nodes during food borne infection of mice. PLoS Pathog 8:e1003015

Chapter 2

Dynamics of Bacteria–Mucus Interactions: Quantitative Assessment of MUC5AC and *Helicobacter pylori* Adhesion to the Gastric Tissue on the Model of *Cavia porcellus*

Weronika Gonciarz and Magdalena Chmiela

Abstract

In this chapter the procedure of immunofluorescence staining of mucin 5 (MUC5AC) in the gastric tissue specimens and primary gastric epithelial cells derived from the *Cavia porcellus* (guinea pig) has been presented in conjunction with the assessment of colonization of gastric tissue with *Helicobacter pylori.*

Key words Mucin 5, Gastric tissue, Primary cells, Cell culture, *H. pylori*, Adhesion

1 Introduction

The gastric mucosa in humans is a physical barrier covered with a mucus layer, which protects stomach against harmful chemicals, enzymes, microbial pathogens, and mechanical disorders [1, 2]. Mucosal mucins are high-molecular glycoproteins about 250–2000 kDa, synthesized by epithelial cells [3]. The sugar chains of mucins are mainly attached to the serine and threonine of the protein core, forming numerous O-glycosidic bonds. Sugars usually constitute over 50% of the mass of mucin (often even 70–80%). Each mucin may contain up to 100 different oligosaccharide residues [4]. A typical terminal modification of oligosaccharide chains is the occurrence of Lewis antigens, which are produced by epithelial cells and classified by function or structure into two groups: transmembrane or secretory [5]. MUC5AC is the dominant component of gastric mucus, which can be used by microorganisms present in the stomach for a temporary or chronic colonization. The Gram-negative bacteria *Helicobacter pylori* (*H. pylori*) colonize the stomach of about 50% people resulting in the development of gastritis and/or duodenitis, gastric or duodenal ulcer, or even gastric cancer or mucosa-associated lymphoid tissue lymphoma

Inka Brockhausen (ed.), *Dynamics of Bacteria-Mucus Interactions*, Methods in Molecular Biology, vol. 2942, https://doi.org/10.1007/978-1-0716-4627-4_2,

[6, 7]. Bacterial flagella facilitate the translocation of these bacteria in the mucus and interaction of bacterial adhesins with MUC5AC which promotes the early step of colonization process, before the later interaction of bacteria with gastric epithelial cells through cell-to-cell interactions. It has been shown that *H. pylori* colocalize with MUC5AC in the human stomach [8]. The glycosylation of the host gastrointestinal tract plays a role in *H. pylori* pathogenesis [9]. Several *H. pylori* adhesins have been identified, including the blood group antigen-binding adhesin A (BabA) and sialic acid-binding adhesin A (SabA), interacting with Lewis b and sialylated Lewis X, respectively [10–12]. Previously we showed that *H. pylori* and their soluble compounds increased the production of MUC5AC and deposition of LeX/Y determinants in the primary gastric epithelial cells of *Cavia porcellus* in vitro in relation to enhanced adhesion of these bacteria to the target cells [13]. These in vitro results were confirmed in *Cavia porcellus* infected experimentally with *H. pylori* [14]. The chronic *H. pylori* infections are thus initiated by the binding of these bacteria to the gastric MUC5AC and are facilitated by the high incidence of on/off mutations in genes encoding *H. pylori* adhesins, including BabA or SabA, minimizing the effectiveness of the immune system, locally [15]. The key medical problem is antibiotic resistance of *H. pylori* isolates to commonly used antibiotics—clarithromycin, metronidazole, levofloxacin, amoxicillin, and tetracyclines [16], resulting in chronic infections and damage of gastric barrier cells. Disintegration of cells can facilitate the translocation of *H. pylori* components and induction of systemic effects [17–19]. The survival strategies of *H. pylori* prompt to search the biologically active components with immunomodulating properties and facilitating the restoration of the gastric cell barrier integrity to restore local homeostasis. The preliminary results show that *Mycobacterium bovis* onco–Bacillus Calmette-Guerin (BCG) bacilli are able to diminish the amount of MUC5AC on the surface of gastric epithelial cells of *Cavia porcellus* in cell cultures in vitro and in vivo during experimental infection, in conjunction with diminishing the binding of *H. pylori* to these cells [14, 17]. Therefore, the quantitative assessment of MUCAC5 together with the imaging of the binding of *H. pylori* to gastric epithelial cells in the *Cavia porcellus* model may be a useful tool in development of the research on the pathogenesis of *H. pylori* infections at the level of gastric tissue. Moreover, the described techniques may be useful in the evaluation of new antibacterial components that limit bacterial colonization (Fig. 1).

All cited experiments involving *Cavia porcellus* (guinea pig) model were developed according to the Animal Research: Reporting of In Vivo Experiments (ARRIVE) guidelines and guidelines and regulations EU directive (Directive 2010/63/EU of the European Parliament and of the Council of 22 September 2010 on the protection of animals used for scientific purposes (Dz.U. L

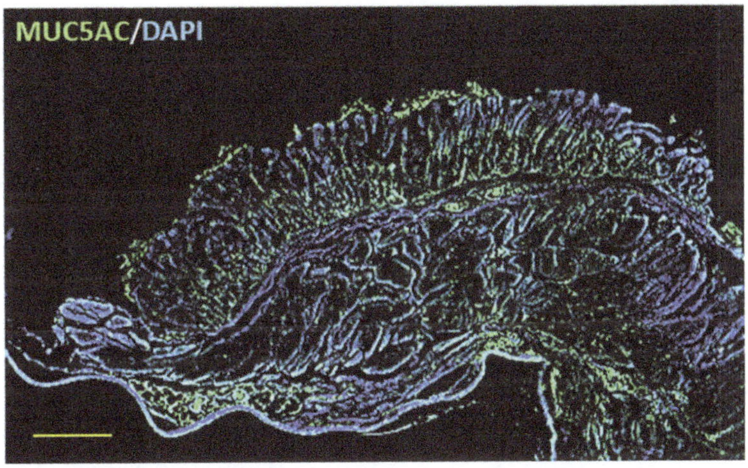

Fig. 1 Representative confocal microscope image of gastric guinea pig tissue sections stained with anti-Mucin 5 (MUC5AC) primary antibodies and secondary fluorescently labeled antibodies. The production of MUC5AC (green) was assessed by staining of tissue section with mouse anti-MUC5AC antibodies and secondary rabbit antibodies against murine immunoglobulins labeled with fluorescein isothiocyanate (FITC). The tissue sections were imaged under a confocal microscope at a wavelength 495 nm (excitation) and 519 nm (emission). Cell nuclei were stained with 4′,6-diamidino-2-phenylindole (DAPI) fluorescent label and visualized at wavelengths 345 nm (excitation) and 455 nm (emission). Barr 100 μm, 10× magnification

276 z 20.10.2010, s. 33–79) and were approved by the Local Ethics Committee (LKE9) for Animal Experiments of the Medical University of Lodz, Poland, which was established by the Ministry of Science and Higher Education in Poland (Ethics Committee decision number: 58/ŁB45/2016).

2 Materials

2.1 Media and Reagents for Cell Cultures and Cell/Tissue Visualization

1. Medium 1: Ham's F-12 medium and DMEM (1:1) supplemented with 10% heat-inactivated FCS (fetal calf serum), 30 min, 56 °C in water bath, 100 U/mL penicillin, 100 U/mL streptomycin, 0.025 μg/mL amphotericin B, L-glutamine 2 mM/mL, 1% HEPES (N-2-hydroxyethylpiperazine-N-2-ethane sulfonic acid), 0.01 μg/mL EGF (epidermal growth factor), and 0.005% dexamethasone solution in Roswell Park Memorial Institute (RPMI)-1640 medium.
2. Solution 1: PBS (phosphate-buffered saline) pH 7.4, liquid, sterile filtered, and suitable for cell culture.
3. Solution 2: 1% BSA (bovine serum albumin) solution in PBS.
4. Solution 3: 5% BSA (bovine serum albumin) solution in PBS.

5. Solution 4: 0.25% trypsin solution in RPMI-1640 medium.
6. Solution 5: 4% formaldehyde solution in PBS.
7. Solution 6: DAPI (4′,6-diamidino-2-phenylindole) solution in PBS.
8. Solution 7: Phalloidin Texas Red fluorescently labeling solution in PBS.
9. Solution 8: primary antibodies against Mucin-5 in *Solution 2* (**item 3**).
10. Solution 9: secondary antibodies fluorescently labeled in *Solution 2* (**item 3**).
11. Solution 10: primary antibodies against *H. pylori* fluorescently labeled in *Solution 2* (**item 3**).
12. Histological (10%) formalin.
13. Ethanol: 70%, 80%, 96%, and 100%.
14. 96-well black cell culture plates with glass bottom.
15. Sterile volumetric pipettes and Pasteur pipettes.
16. Inverted light microscope.
17. Confocal microscope.
18. Glass coverslip.
19. Cell culture incubator with controlled temperature, humidity, and CO_2 atmosphere.

3 Methods

3.1 Isolation and Culture of Guinea Pig Primary Gastric Tissue Cells

Isolate the stomach from the euthanized animal (overdose of sodium barbiturate), according to the Animal Research: Reporting of In Vivo Experiments (ARRIVE) guidelines and guidelines and regulations of EU directive (Directive 2010/63/EU of the European Parliament and of the Council of 22 September 2010 on the protection of animals used for scientific purposes [19]. A detailed step-by-step method for cell isolation and cultivation has been described previously [20].

3.2 Preparation of Gastric Tissue for Staining

1. Isolate the stomach from the euthanized animal and prepare tissue as previously described [20].
2. Cut the gastric tissue into small pieces of approximately 1 cm.
3. Fix the tissue in histological (10%) formalin.
4. Dehydrate the tissue.
5. Fix the tissue in paraffin.
6. Cut the tissue on a microtome into thin sections up to 4 μm.
7. Deparaffinize the specimens.

8. Dehydrate the tissue by immersing it in a descending series of ethanol concentrations (70% ethanol, 80% ethanol, 96% ethanol, and 100% ethanol).
9. Wash the tissue 15 min in distilled water.

3.3 Imaging of MUC5AC in Guinea Pig Primary Gastric Cells in a Confocal Microscope (Fig. 2)

1. Culture of gastric primary cells derived from guinea pig in *Medium 1* (*see* **Notes 1–3**).
2. Bring the cell suspension to a density of 2×10^4 cells/mL in *Medium 1*.
3. Distribute 100 mL of cell suspension into the wells of 96-well cell culture glass plates with the black bottom.
4. Incubate the cell suspension overnight (5% CO_2, 37 °C).
5. Look at the cells under the inverted light microscope to control their morphology.

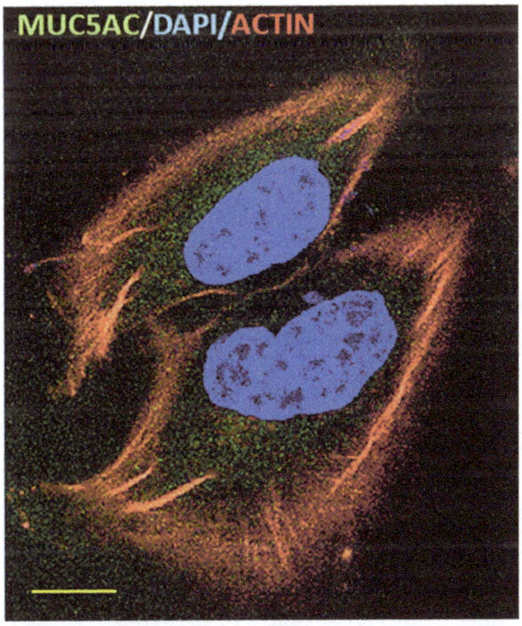

Fig. 2 Representative image of guinea pig primary gastric cells developed in a confocal microscope after the immunostaining with anti-mucin 5 (MUC5AC) antibodies and secondary fluorescently labeled antibodies. MUC5AC (green) was assessed by staining of cells with primary mouse anti-MUC5AC antibodies and secondary rabbit antibodies against murine immunoglobulins labeled with fluorescein isothiocyanate (FITC). The cell imaging was performed using a confocal microscope at wavelength 495 nm (excitation) and 519 nm (emission). Cell nuclei (blue) were stained with 4′,6-diamidino-2-phenylindole (DAPI) fluorescent label and visualized at wavelengths 345 nm (excitation), and 455 nm (emission). Actin (red) was stained with Texas Red phalloidin fluorescent label and visualized at wavelengths 591 nm (excitation) and 608 nm (emission). Barr 10 μm, 64× magnification

6. Remove *Medium 1* from the plate wells.
7. Wash the cell monolayer once with 100 μL *Solution 1* (*see* **Note 4**).
8. Fix the cells with 100 μL *Solution 5*.
9. Incubate 15 min (room temperature).
10. Remove *Solution 5* from the wells.
11. Wash the cell monolayer three times with 100 μL *Solution 1* (*see* **Note 4**).
12. Add to the wells 300 μL *Solution 3* (*see* **Note 5**).
13. Leave the plates for 60 min at room temperature.
14. Add 100 μL *Solution 8*.
15. Incubate overnight (at 4 °C).
16. Remove *Solution 8* from the wells.
17. Wash the cell monolayer three times with 100 μL *Solution 1* (*see* **Note 5**).
18. Add 100 μL *Solution 9*.
19. Leave the plates at room temperature for 120 min in the dark.
20. Remove *Solution 9* from the wells.
21. Wash the cell monolayer three times with 100 μL *Solution 1* (*see* **Note 5**).
22. Add 100 μL *Solution 6* to all wells (*see* **Note 6**).
23. Leave the plate at room temperature in the dark for 20 min (*see* **Note 6**).
24. Remove *Solution 6* from the wells (*see* **Note 7**).
25. Wash the cell monolayer three times with 100 μL *Solution 1* (*see* **Note 5**).
26. Add 100 μL Solution 7 to all wells (*see* **Note 6**).
27. Leave the plate at room temperature in the dark for 20 min (*see* **Note 6**).
28. Remove *Solution 7* from the wells.
29. Wash the cell monolayer three times with 100 μL *Solution 1* (*see* **Note 5**).
30. Add 100 μL Solution 1 to all wells (*see* **Note 6**).
31. Visualize the cells in a confocal microscope under induction and excitation wavelength conditions dedicated to specific dyes (*see* **Note 8**).

3.4 Imaging of MUC5AC in the Guinea Pig Gastric Tissue Sections in a Confocal Microscope

1. Immerse the gastric tissues in the *Solution 3*.
2. Incubate 60 min (room temperature).
3. Apply 200 μL *Solution 8* to each tissue section.
4. Incubate overnight (at 4 °C), in the humid chamber.
5. Remove *Solution 8* from the tissue section.
6. Wash the tissue section three times with *Solution 1* (*see* **Note 5**).
7. Apply 200 μL *Solution 9* to each tissue section.
8. Incubate at room temperature in the dark for 120 min, in the humid chamber (*see* **Note 6).**
9. Remove *Solution 9* from each tissue section.
10. Wash each tissue section three times with *Solution 1* (*see* **Note 5**).
11. Apply 200 μL *Solution 6* to each tissue section (*see* **Note 6**).
12. Incubate at room temperature in the dark for 20 min (*see* **Note 6**).
13. Remove *Solution 6* from each tissue section.
14. Wash each tissue section three times with *Solution 1* (*see* **Note 5**).
15. Air-dry the tissue preparations.
16. Close tissue sections in fixation balm (*see* **Note 8**).
17. Cover with a coverslip.
18. Visualize tissue sections in a confocal microscope under induction and excitation wavelength conditions dedicated to specific dyes (*see* **Note 9**).

3.5 Assessment of Helicobacter pylori Adhesion to Guinea Pig Primary Gastric Cells by the Immune Staining and Imaging in a Confocal Microscope (Fig. 3)

1. Culture primary cells with *Medium 1* (*see* **Notes 1–3**).
2. Bring to a density of 2×10^4 cells/mL in *Medium 1*.
3. Add 100 mL of cell suspension in a well of a 96-well glass-bottomed black plate.
4. Incubate overnight (5% CO_2, 37 °C).
5. Look at the cells under the microscope, and assess their appearance.
6. Remove *Medium 1* from the plate wells.
7. Wash the monolayer once with 100 μL *Solution 1* (*see* **Note 4**).
8. Add live *Helicobacter pylori* suspension (multiplicity of infection (MOI): 10:1).
9. Incubate 2 h (5% CO_2, 37 °C).
10. Wash the monolayer once with 100 μL *Solution 1* (*see* **Note 4**).
11. Fix the cells with 100 μL *Solution 5*.
12. Leave the cells for 15 min at room temperature.

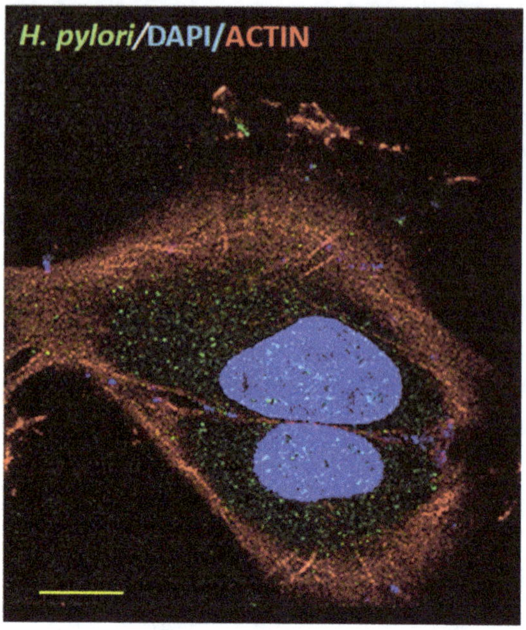

Fig. 3 Representative image of guinea pig primary gastric cells with attached *Helicobacter pylori* in a confocal microscope. Binding of *H. pylori* to gastric epithelial cells was assessed by staining with anti-*H. pylori* antibodies labeled with fluorescein isothiocyanate (FITC). The cell imaging for *H. pylori* (green) was performed using a confocal microscope at wavelength 495 nm (excitation) and 519 nm (emission). Cell nuclei (blue) were stained with 4′,6-diamidino-2-phenylindole (DAPI) fluorescent label and visualized at wavelengths 345 nm (excitation) and 455 nm (emission). Actin (red) was stained with Texas Red phalloidin fluorescent label and visualized at wavelengths 591 nm (excitation) and 608 nm (emission). Barr 100 μm, 64× magnification

13. Remove *Solution 5* from the wells.
14. Wash the cell monolayer three times with 100 μL *Solution 1* (*see* **Note 4**).
15. Add 300 μL *Solution 2* (*see* **Note 5**).
16. Incubate 60 min at room temperature.
17. Remove *Solution 2* from the wells.
18. Add 100 μL *Solution 10.*
19. Incubate the cells overnight at 4 °C.
20. Remove *Solution 10* from the wells.
21. Wash the cell monolayer three times with 100 μL *Solution 1* (*see* **Note 5**).
22. Add 100 μL *Solution 6* to all wells (*see* **Note 6**).
23. Leave the plates at room temperature in the dark for 20 min (*see* **Note 6**).

24. Remove *Solution 6* from the wells.
25. Wash the cell monolayer three times with 100 μL *Solution 1* (*see* **Note 5**).
26. Add 100 μL *Solution 7* to all wells (*see* **Note 6**).
27. Incubate the plates at room temperature in the dark for 20 min (*see* **Note 6**).
28. Remove *Solution 7* from the wells.
29. Wash the cell monolayer three times with 100 μL *Solution 1* (*see* **Note 5**).
30. Add 100 μL Solution 1 to all the wells (*see* **Note 6**).
31. Visualize the cells exposed to *H. pylori* in a confocal microscope under induction and excitation wavelengths conditions dedicated to specific dyes used for labeling of secondary antibodies (*see* **Note 9**).

4 Notes

1. Remember that the number of cell cycles of primary cells is limited. Check the cell morphology and cell viability by trypan blue exclusion assay prior to each experiment.
2. The primary cells are more sensitive than neoplastic cells to experimental conditions, so handle them carefully particularly when washing. This will prevent the cell monolayer from peeling off before further specific treatment.
3. If you are not currently using cells, store them in −80°C in sterile freezing tubes. Prepare stocks in standard freezing medium (mix 1 mL cell culture with dimethyl sulfoxide 10% and 9 mL heat-inactivated (FCS)). Grow cells from the stock for the next experiment. Total number of cell passages is 10.
4. When you rinse the cells, be careful not to damage the cell monolayer.
5. Remember to thoroughly and completely fill the wells with blocking buffer to avoid nonspecific binding of antibodies.
6. For visualizing selected cell organelles, use DAPI and Phalloidin Texas Red fluorescent labels in a concentration recommended by the manufacturer's recommendation.
7. Carry out with DAPI and/or Phalloidin Texas Red within the time specified by the manufacturer.
8. Before imaging the tissue sections in a confocal microscope, remember to wait until the balm has thoroughly fixed the tissue and the coverslip has adhered thoroughly, otherwise the preparation may be damaged.

9. Visualize cells in a confocal microscope immediately after staining, or after a maximum of 7 days as the dyes bleach. If you cannot visualize the cells immediately, keep the stained plates in the refrigerator (4 °C).

References

1. Tarnawski A (1998) Cellular and molecular mechanisms of mucosal defense and repair. In: Bioregulation and its disorders in gastrointestinal tract. Blackwell Science Japan, Tokyo, pp 3–7.
2. Amieva MR, El-Omar EM (2008) Host-bacterial interactions in *Helicobacter pylori* infection. Gastroenterology 134:306–323. https://doi.org/10.1053/j.gastro.2007.11.009
3. Song C, Chai Z, Chen S et al (2003) Intestinal mucus components and secretion mechanisms: what we do and do not know. Exp Mol Med 55:681–691. https://doi.org/10.1038/s12276-023-00960-y
4. Robbe C, Capon C, Coddeville B et al (2004) Structural diversity and specific distribution of O-glycans in normal human mucins along the intestinal tract. Biochem J 384:307–316. https://doi.org/10.1042/BJ20040605
5. Derrien M, van Passel MW, van de Bovenkamp JH et al (2010) Mucin-bacterial interactions in the human oral cavity and digestive tract. Gut Microbes 1:254–268. https://doi.org/10.4161/gmic.1.4.12778.
6. Malfertheiner P, Camargo MC, El-Omar E et al (2023) *Helicobacter pylori* infection. Nat Rev Dis Prim 9:19. https://doi.org/10.1038/s41572-023-00431-8
7. Sukri A, Hanafiah A, Mohamad Zin N et al (2020) Epidemiology and role of *Helicobacter pylori* virulence factors in gastric cancer carcinogenesis. APMIS 128:150–161. https://doi.org/10.1111/apm.13034
8. Van den Brink GR, Tytgat KM, Van der Hulst RW et al (2000) *H. pylori* colocalizes with MUC5AC in the human stomach. Gut 46:601–607. https://doi.org/10.1136/gut.46.5.601
9. Moran AP, Gupta A, Joshi L (2011) Weet-talk: role of host glycosylation in bacterial pathogenesis of the gastrointestinal tract. Gut 60:1412–1425. https://doi.org/10.1136/gut.2010.212704
10. Borén T, Falk P, Roth KA et al (1993) Attachment of *Helicobacter pylori* to human gastric epithelium mediated by blood group antigens. Science 262:1892–1895. https://doi.org/10.1126/science.8018146
11. Mahdavi J, Sondén B, Hurtig M et al (2002) *Helicobacter pylori* SabA adhesin in persistent infection and chronic inflammation. Science 297:573–578. https://doi.org/10.1126/science.1069076
12. Ishijima N, Suzuki M, Ashida H et al (2011) BabA-mediated adherence is a potentiator of the *Helicobacter pylori* type IV secretion system activity. J Biol Chem 286:25256–25264. https://doi.org/10.1074/jbc.M111.233601
13. Gonciarz W, Walencka M, Moran AP et al (2019) Upregulation of MUC5AC production and deposition of Lewis determinants by *Helicobacter pylori* facilitate gastric tissue colonization and the maintenance of infection. J Biomed Sci 26:23. https://doi.org/10.1186/s12929-019-0515-z
14. Gonciarz W, Chyb M, Chmiela M (2023) Diminishing of *Helicobacter pylori* adhesion to *Cavia porcellus* gastric epithelial cells by BCG vaccine mycobacteria. Sci Rep 13:16305. https://doi.org/10.1038/s41598-023-43571-3
15. Doohan D, Rezkitha YAA, Waskito LA et al (2021) *Helicobacter pylori* BabA-SabA key roles in the adherence phase: the synergic mechanism for successful colonization and disease development. Toxins 13:485. https://doi.org/10.3390/toxins13070485
16. Nestergard O, Moayeri B, Halvorsen FA et al (2022) *Helicobacter pylori* resistance to antibiotics before and after treatment: incidence of eradication failure. PLoS One 17:e0265322. https://doi.org/10.1371/journal.pone.0265322
17. Gonciarz W, Płoszaj P, Chmiela M (2024) *Mycobacterium bovis* BCG reverses deleterious effects of *H. pylori* components towards gastric barrier cells *in vitro*. Biomed Pharmacother 178:117193. https://doi.org/10.1016/j.biopha.2024.117193
18. Mnich E, Kowalewicz-Kulbat M, Sicińska P et al (2016) Impact of *Helicobacter pylori* on the healing process of the gastric barrier. World J Gastroenterol 22:7536–7558. https://doi.org/10.3748/wjg.v22.i33.7536
19. Gonciarz W, Krupa A, Hinc K et al (2019) The effect of *helicobacter pylori* infection and different *H. pylori* components on the proliferation

and apoptosis of gastric epithelial cells and fibroblasts. PLoS One 7:e0220636. https://doi.org/10.1371/journal.pone.0220636

20. Gonciarz W, Chmiela M (2024) Isolation, culture, and microscopic imaging of guinea pig primary gastric tissue cells. In: Baratta M (ed) Epithelial cell culture, Methods in molecular biology, vol 2749. Humana, New York. https://doi.org/10.1007/978-1-0716-3609-1_6

Chapter 3

Growth and Polarization of A549 Human Alveolar Epithelial Cells as a Model to Study Biofilm Formation in *Acinetobacter baumannii*

Luis A. Ramos, Elizabeth C. Walls, Mia S. Petrucci, William F. Penwell, Emily J. Ohneck, Luis A. Actis, and Steven E. Fiester

Abstract

Acinetobacter baumannii is an opportunistic pathogen that has a propensity to cause infections in immunocompromised patients. With the increasing incidence of antibiotic-resistant isolates, *Acinetobacter baumannii* has emerged as one of the most problematic pathogens in recent decades with the World Health Organization urging prioritization for additional research focusing on this nosocomial pathogen. The following protocol outlines the process of culturing and polarizing A549 human alveolar epithelial cells for infection with *Acinetobacter baumannii* to study biofilm formation. The protocol additionally outlines the process of fixing the infected A549 cells for downstream applications.

Key words A549 human alveolar epithelial cells, Cell polarization, *Acinetobacter*, Respiratory infection model, Biofilm formation

1 Introduction

Gram-negative *Acinetobacter baumannii* is an opportunistic pathogen primarily affecting immunocompromised patients causing urinary and systemic infections, secondary meningitis, nosocomial burn infections, pneumonia, and necrotizing fasciitis. *A. baumannii* infections make up about 10% of infections found in intensive care units and, since 2008, have been listed as the fifth most common Gram-negative nosocomial pathogen [1, 2]. *A. baumannii*'s virulence factors including iron acquisition systems, hemolytic factors, biofilm-associated virulence factors, pili, efflux pumps, and antibiotic resistance gene products result in severe infections that are difficult to treat. With its resistance to a wide range of antibiotics and ability to respond to the harsh host environmental conditions it encounters during colonization and

Inka Brockhausen (ed.), *Dynamics of Bacteria-Mucus Interactions*, Methods in Molecular Biology, vol. 2942, https://doi.org/10.1007/978-1-0716-4627-4_3,

infection, including those associated with iron depletion and the generation of oxygen radicals, *A. baumannii* is of significant concern for public health globally. An understanding of the virulence mechanisms of *A. baumannii* is therefore critically needed to inform the development of novel therapeutics to treat this pathogen. Many *Acinetobacter* studies have focused on the elucidation of resistance mechanisms; however, few have investigated the virulence mechanisms of this pathogen making the need for methodologies that investigate virulence traits such as biofilm formation of paramount importance.

Bacterial biofilms are formed by the adhesion of bacterial cells to substrates, surfaces, and/or each other and are surrounded by a polymeric matrix creating a defense mechanism that increases survival for the respective bacterial cells [2, 3]. The process of bacterial biofilm formation is controlled by regulatory pathways that are impacted by the presence or absence of antibiotics, bacterial cell density, and/or environmental conditions. Biofilm-producing bacteria therefore have the ability to survive harsh environmental conditions such as those found in the human body where these bacteria evade the host immune system. This response results in increased antibiotic resistance and the pathogenicity of serious infections [4]. Biofilm assays provide insight on adherence mechanisms, biofilm structure, and the roles of biofilm as a virulence factor.

In vitro cell culturing methods are beneficial for mimicking host–microbe interactions in a controlled environment and allow for the study of bacterial biofilm formation and cytotoxicity in the presence of host cells. In vitro propagation of immortalized cells does however require a complex mixture of additives to enhance cell growth and ensure proliferation. Dulbecco's Modified Eagle Media (DMEM), composed of low concentrations of glucose (1 g/L) and sodium pyruvate, is a common basal media broadly used for the culturing of adherent mammalian, plant, and insect cells [5]. Various forms of DMEM can be found including those with different concentrations of glucose to support cell lines with sensitivity to sugars. The original medium was produced in the 1950s and, with its lack of nutrients, was unable to sustain cells on its own [6]. Presently, the medium is often supplemented with amino acids, fetal bovine serum (FBS), and other nutrients needed to support growth. A complete DMEM medium, as is used in this procedure, also includes FBS to support cell attachment and growth, as well as L-glutamine used by cells in the synthesis of proteins and nucleic acids. Since the cultures described in this method were incubated in CO_2, sodium bicarbonate was added to control the pH of the cultures and ensure they remained neutral. Throughout this protocol, it is essential to continuously remove and replace the DMEM media from each subculture to replenish nutrients utilized by cells during growth.

Transwell permeable supports for use in cell culture plates are typically used to study processes such as solute transport across a cell monolayer while minimizing the handling and potential contamination of cell cultures. A transwell permeable support, often referred to as a transwell insert, is a tapered cylinder with a lip allowing suspension of the transwell insert within a well of a cell culture plate and a permeable membrane at the base of the transwell insert on which cells can be propagated. Since the transwell insert is suspended within the well of a cell culture plate, the permeable membrane can either be completely submerged in culture medium allowing cells grown on the membrane to be exposed to the culture medium on both the apical and basal surfaces or partially submerged wherein only the basal surface of cells would be exposed to culture media while the apical surface is exposed to unsubmerged ambient conditions. The later condition is an environment in which cells, such as A549 human alveolar epithelial cells, can not only divide and form a monolayer on the permeable membrane but also polarize to produce mucins and morphologically mimic in vivo alveolar epithelial cells when unsubmerged. Once polarized, A549 cells can be infected with *A. baumannii* to study host–pathogen interactions such as biofilm formation. Following infection, cellular fixation with formaldehyde in Hank's Balanced Salt Solution (HBSS) is used to conserve the bacterial biofilm, mucins, and underlying A549 cells for further examination with techniques such as scanning electron microcopy [7]. This process has been performed repeatedly and has resulted in several high-quality images allowing visualization of *A. baumannii* biofilm formation on the surface of polarized A549 human alveolar epithelial cells [2, 8–12].

2 Materials

All buffers, media, and solutions should be obtained from a water purification system producing ultrapure water with a sensitivity of 18 MΩ-cm at 25 °C and stored at room temperature unless otherwise indicated. All chemicals used in the preparation of buffers and solutions should be reagent grade at minimum, unless indicated otherwise. The disposal of any generated waste should follow institutional guidelines as well as local, state, and federal statutes.

2.1 Cell Culture

1. Cryogenic tubes.
2. A549 human alveolar epithelial cells (ATCC CCL-185).
3. Dimethylsulfoxide (DMSO), cell culture grade.
4. Liquid nitrogen dewar filled with liquid nitrogen.
5. Float rack.

6. General purpose laboratory water bath capable of 37 °C.
7. Microcentrifuge capable of at least 5000×*g*.
8. 100–1000 μL micropipette with sterile micropipette tips.
9. Refrigerator capable of 4 °C.
10. Dulbecco's Modified Eagle Medium (DMEM) complete medium: DMEM with 10% fetal bovine serum. Store at 4 °C.
11. Motorized pipetting controller.
12. 5 mL serological pipettes.
13. T25 tissue culture flask.
14. CO_2 cell culture incubator capable of 37 °C and humidification.
15. Inverted microscope.
16. 10 mL serological pipettes.
17. Sterile-filtered phosphate-buffered saline (PBS): 137 mM NaCl, 2.7 mM KCl, 10 mM Na_2HPO_4, 2 mM KH_2PO_4, pH 7.4.
18. Trypsin-EDTA. Store at −20°.
19. 15 mL conical centrifuge tubes.
20. Benchtop centrifuge capable of 500×*g* with rotor for 15 mL conical tubes.
21. 2 mL microcentrifuge tubes, sterile.
22. Trypan blue: 0.4% solution.
23. 1–10 μL micropipette with sterile micropipette tips.
24. Glass coverslips.
25. Hemocytometer.
26. Ethyl alcohol: 70% solution in water.
27. Corning® HTS Transwell®-24 well permeable supports (Catalog Number: CLS3396; Corning, NY).
28. Sterile forceps.
29. Class II type A2 biological safety cabinet.

2.2 Bacteriological Culture

1. 100 mm × 15 mm sterile Petri dishes.
2. *Acinetobacter baumannii* strain ATCC 19606^T [81, DSM 6974].
3. −80 °C freezer.
4. Luria-Bertani, Miller agar (LBA).
5. Sterile plastic inoculating loops.
6. Microbiological incubator capable of 37 °C.
7. Sterile culture tubes.

8. Luria-Bertani, Miller broth (LB).
9. Bacterial shaker incubator capable of shaking at 200 rpm at 37 ° C.
10. Visible wavelength spectrophotometer.
11. Spectrophotometer microcuvettes.
12. Hank's Balanced Salt Solution (HBSS): 140 mM NaCl, 5 mM KCl, 1 mM $CaCl_2$, 0.4 mM $MgSO_4 \cdot 7\ H_2O$, 0.5 $MgCl_2 \cdot 6\ H_2O$, 0.3 mM $Na_2HPO_4 \cdot 2\ H_2O$, 0.4 mM KH_2PO_4, 6 mM D-glucose, 4 mM $NaHCO_3$, pH 7.4.

2.3 Sample Fixation

1. Fixation solution: 4% formaldehyde solution in HBSS. Store at 4 °C.
2. Sterile scalpel.

3 Methods

Unless otherwise specified, all steps should be performed at room temperature using aseptic techniques in a class II type A2 biological safety cabinet (*see* **Note 1**). All procedures should be performed under biological safety level 2 conditions.

3.1 Preparation of A549 Cell Lines

1. Remove a cryovial of 10^6 A549 human alveolar epithelial cells/ mL stock culture, suspended in DMEM with 10% DMSO as a cryoprotectant, from liquid nitrogen storage using appropriate protective equipment (*see* **Note 2**).
2. Place the cryovial into a floating rack in a 37 °C general purpose laboratory water bath for 2 min for thawing.
3. Spin the cryovial in a centrifuge at 500×*g* for 2 min.
4. Aspirate the media from the vial using a sterile pipette.
5. Gently resuspend the cells in 1 mL of fresh DMEM complete medium (*see* **Note 3**).
6. Use sterile pipettes to add 1 mL of the cell suspension and 5 mL of fresh DMEM complete medium to a Corning® T25 cell culture flask (*see* **Note 4**).
7. Incubate the cell culture flask in a 37° CO_2 incubator overnight.
8. Check cell confluency every 24 h using an inverted microscope.
9. Media in the cell culture flask should be aspirated and replaced every 48 h with 6 mL of fresh DMEM complete medium using a sterile serological pipette until confluency is above 90%.
10. Once cell confluency is 90% or greater, aspirate the medium from the A549 T25 flask with a sterile serological pipette.

11. Pipette 2 mL of PBS into the flask, gently agitate and remove PBS and any remaining medium with the pipette.
12. Add 1 mL trypsin EDTA and place the flask in the 37° CO_2 incubator for 5 min.
13. Once cell detachment is visually confirmed under an inverted microscope, add 5 mL of fresh DMEM complete medium.
14. Pipette cell suspension into a sterile 15 mL conical tube.
15. Centrifuge at 500×*g* for 2 min.
16. Remove supernatant using a sterile serological pipette and resuspend cells in 6 mL of fresh medium.
17. Collect 100 μL of cell suspension using a micropipette and place the suspension in a sterile microcentrifuge tube.
18. Add 100 μL of 0.4% trypan blue to the microcentrifuge tube, and gently invert the tube twice to mix the dye.
19. Add 10 μL into each end of a hemocytometer chamber with a coverslip previously cleaned with alcohol and place the hemocytometer under the compound light microscope to count cells.
20. If cell density is less than 10^5 cells/mL, add the cell suspension in the 15 mL conical tube into a T25 flask, place the flask back into the CO_2 incubator at 37 °C, and repeat **steps 8–19**.

    Once a cell density of 10^5 cells/mL has been reached, proceed to Subheading 3.2.

3.2 Preparation of Transwell Support Membranes

The following can be repeated for the number of transwell inserts the specific investigator's experiments require.

1. Use a micropipette to add 600 μL of fresh DMEM complete medium into a well of a 24-well microtiter plate included with Corning® HTS Transwell®-24 well permeable supports.
2. Use sterile forceps to place a transwell insert into the well containing DMEM complete medium.
3. Add 100 μL of the A549 cell suspension prepared in **step 14** of Subheading 3.1 to the center of the transwell insert using a micropipette.
4. Add 400 μL of fresh DMEM complete medium into the transwell insert using a micropipette.

3.3 Maintenance of Transwell Inserts

1. Exchange DMEM complete medium on both the apical and basal surfaces of the permeable membrane every 48 h until 100% confluency is confirmed visually using an inverted microscope (*see* **Note 5**).

2. Once 100% confluency is reached, replace the DMEM medium on only the basal surface of the permeable membrane every 48 h for A549 cell polarization to occur (*see* **Note 6**).
3. A cell count should be performed every 48 h to check for cell density and viability by aspirating the medium from one well with a sterile micropipette, adding 1 mL of trypsin and incubating at 22 °C for 1 h. One milliliter of fresh DMEM medium is then added to the well, and Steps 17–19 in Subheading 3.1 are repeated (*see* **Note 7**).

3.4 Preparation of Bacterial Culture

1. Prepare a quadrant streak of *A. baumannii* from a −80 °C frozen stock onto a Luria-Bertani agar (LBA) plate using a sterile inoculating loop and incubate in a microbiological incubator for 16 h at 37 °C (*see* **Note 1**).
2. Use a sterile inoculating loop to transfer a single colony from the LBA plate into a culture tube containing 1 mL of Luria-Bertani (LB) broth and incubate in a bacterial shaker incubator for 16 h at 37 °C with shaking at 200 rpm.
3. Power on a spectrophotometer, and allow it to warm up for several minutes.
4. Aseptically transfer 50 μL of bacterial culture and 450 μL of HBSS into a sterile microcentrifuge tube and agitate gently to obtain a 1:10 culture dilution.
5. Transfer the 500 μL 1:10 culture dilution and 500 μL of HBSS into two individual microcuvettes.
6. Set the wavelength of the spectrophotometer to 600 nm, and use the cuvette containing 500 μL of HBSS as a blank.
7. Place the cuvette containing the 1:10 culture dilution into the spectrophotometer to obtain an OD_{600} measurement.
8. Multiply the OD_{600} measurement by ten to account for the dilution and use a previously made growth curve to obtain the CFU/mL of the culture.
9. Use the OD_{600} measurement and the equation concentration (initial) × volume (initial) = concentration (final) × volume (final) to calculate the volume of culture needed to make a 1 mL solution of 1×10^5 CFU/mL, and use a micropipette with sterile tips to make this solution in a sterile microcentrifuge tube.
10. Centrifuge the microcentrifuge tube at 5000×*g*, use a sterile micropipette to remove supernatant and resuspend the pellet in 1 mL of sterile HBSS.

3.5 Infection

1. Use a micropipette and a sterile pipette tip to transfer eight, 1-μL aliquots of the diluted 1×10^5 CFU/mL culture onto the polarized A549 cells making certain each drop is placed in a different location.
2. Exchange the DMEM medium on only the basal surface of the permeable membrane (*see* **Note 8**).
3. Place the tissue culture plate in a CO_2 incubator at 37 °C for 24 h.

3.6 Sample Fixation for Downstream Applications

1. Aseptically aspirate the medium from the basal surface of the permeable membrane.
2. Add 1 mL of HBSS to the basal surface of the permeable membrane to remove residual DMEM, and then aspirate the HBSS.
3. Use a serological pipette to add 2 mL of fixation solution to the basal surface of the permeable membrane and 1 mL of fixation solution to the apical surface of the permeable membrane.
4. Incubate the plate at 4 °C for 24 h (*see* **Note 9**).

4 Notes

1. *Acinetobacter baumannii* ATCC 19606^T requires biosafety level 2 precautions and therefore manipulation of this organism must occur in a biological safety cabinet. While we recommend a class II type A2 biological safety cabinet, a class II type B2 biological safety cabinet or equivalent would also be acceptable. As a biosafety level 2 bacterium, *A. baumannii* will require approval from the institution at which the work is to be completed most likely in the form of an approved institutional biosafety committee protocol and authorization from the institution's biological safety officer.
2. Personal protective equipment (PPE) for handling liquid nitrogen should include cryogenic gloves, a face shield, and an apron at minimum as well as any other institutionally mandated PPE.
3. Since DMSO is an organic solvent, the exchange of DMEM containing 10% DMSO for DMEM free of DMSO should be performed as quickly as possible to prevent cellular damage.
4. During preparation of A549 cells (Subheading 3.1), DMEM can be supplemented with penicillin-streptomycin at a concentration of 50–100 I.U./mL to mitigate contamination. While this step is unnecessary, it is worth consideration if contamination becomes an issue. Penicillin-streptomycin should never be added to transwell inserts as the antibiotics may impact bacterial physiology and/or kill the cells.

5. The transwell inserts have three openings for serological pipette tips to fit through for sterile aspiration and replacement of old media.
6. One hundred percent confluency usually occurs after 2 weeks of growth. The polarized cells should then be maintained for at least 12 days before infection. It is imperative to not replace media on the apical surface of the membrane from this step forward. If medium is added to the apical surface, the A549 cells will not polarize.
7. For cost savings, cell cultures can be maintained in standard wells of a 24-well plate in parallel with the transwell inserts. The cells maintained in the wells without transwell inserts can be used to assess cell density and viability in lieu of expending transwell inserts.
8. Do not replace media on the apical surface of the membrane during infection. While the A549 cells have polarized at this point, bacteria suspended in media will behave differently than those exposed to ambient conditions such as those found in an infection.
9. A sterile scalpel may be used to extract the permeable membrane from the transwell insert for downstream processing.

Acknowledgments

No financial interests to disclose.

References

1. Fiester SE, Actis LA (2013) Stress responses in the opportunistic pathogen *Acinetobacter baumannii*. Future Microbiol 8(3):353–365
2. Ohneck EJ, Arivett BA, Fiester SE et al (2018) Mucin acts as a nutrient source and a signal for the differential expression of genes coding for cellular processes and virulence factors in *Acinetobacter baumannii*. PLoS One 13(1): e0190599
3. Zhao A, Sun J, Liu Y (2023) Understanding bacterial biofilms: from definition to treatment strategies. Front Cell Infect Microbiol 13: 1137947
4. Vestby LK, Grønseth T, Simm R et al (2020) Bacterial biofilm and its role in the pathogenesis of disease. Antibiotics 9(2):59
5. Grossner T, Haberkorn U, Hofmann J et al (2022) Effects of different basal cell culture media upon the osteogenic response of hMSCs evaluated by 99mTc-HDP labeling. Int J Mol Sci 23(11):6288
6. Yao T, Asayama Y (2017) Animal-cell culture media: history, characteristics, and current issues. Reprod Med Biol 16(2):99–117
7. Kim S-O, Kim J, Okajima T et al (2017) Mechanical properties of paraformaldehyde-treated individual cells investigated by atomic force microscopy and scanning ion conductance microscopy. Nano Converg 4:1–8
8. Fiester SE, Arivett BA, Beckett AC et al (2019) Miltefosine reduces the cytolytic activity and virulence of *Acinetobacter baumannii*. Antimicrob Agents Chemother 63(1):e01409–e01418
9. Wood CR, Ohneck EJ, Edelmann RE et al (2018) A light-regulated type I pilus contributes to *Acinetobacter baumannii* biofilm, motility, and virulence functions. Infect Immun 86(9). https://doi.org/10.1128/iai.00442-00418
10. Álvarez-Fraga L, Pérez A, Rumbo-Feal S et al (2016) Analysis of the role of the

LH92_11085 gene of a biofilm hyper-producing *Acinetobacter baumannii* strain on biofilm formation and attachment to eukaryotic cells. Virulence 7(4):443–455

11. Rumbo-Feal S, Perez A, Ramelot TA et al (2017) Contribution of the *A. baumannii* A1S_0114 gene to the interaction with eukaryotic cells and virulence. Front Cell Infect Microbiol 7:108
12. Alvarez-Fraga L, Rumbo-Feal S, Perez A et al (2017) Global assessment of small RNAs reveals a non-coding transcript involved in biofilm formation and attachment in *Acinetobacter baumannii* ATCC 17978. PLoS One 12(8): e0182084

Chapter 4

Mucins as Inhibitors Against Bacterial Adhesion to Epithelial Cells

Licínia Santos and Sara K. Lindén

Abstract

The mucus layer protects the epithelium and is a barrier against pathogens. Mucins, a component of the mucus, protect the epithelial surfaces from pathogens by binding and removing them. Different cell lines have been used to study the adhesion of bacteria to epithelial cells and the downstream effects of interactions between pathogens and epithelia. Here, we present a method to evaluate the ability of mucins to inhibit the binding of bacteria to epithelial cells. For this example, we used gastric mucins from a noninfected human, the bacteria *Helicobacter pylori* strain P12, and the gastric adenocarcinoma cell line MKN7. However, the method also works with mucins from other sources and other bacteria and cell lines. Considerations for choosing appropriate cell lines and how to adapt the protocol for fast-growing bacteria are also discussed.

Key words Adhesion, Decoy, Epithelial cells, *H. pylori*, Mucins, Mucus

1 Introduction

The mucus layer is a barrier against pathogens; however, it is also a niche in which bacteria can survive and proliferate [1–3]. The gastric pathogen *Helicobacter pylori* can be found both attached to gastric epithelial cells and residing in the mucus layer. When bound to epithelial cells, *H. pylori* can employ virulence factors and induce inflammatory responses [4, 5].

The mucosal surfaces are sites where mucins and pathogens interact, with the former playing key roles in the protection against infection. Mucins can bind and remove pathogens such as *H. pylori* and can also regulate pathogen virulence, quorum sensing, and growth [6–9]. It has been demonstrated that rhesus monkeys that produce mucins that bind *H. pylori* in a more effective manner have a reduced *H. pylori* density in their stomachs [10]. Furthermore, mice lacking the Muc5AC or Muc1 mucins present a higher gastric density of *H. pylori* after experimental infection than wild-type mice

Inka Brockhausen (ed.), *Dynamics of Bacteria-Mucus Interactions*, Methods in Molecular Biology, vol. 2942,
https://doi.org/10.1007/978-1-0716-4627-4_4,

[11, 12]. Additionally, Muc1 can act as a releasable decoy, limiting *H. pylori* access to gastric epithelial cells [12]. This suggests that the mucins' ability to bind to *H. pylori* can impair its adhesion to gastric epithelial cells.

Gastric adenocarcinoma cell lines, such as MKN7, are often used to study *H. pylori* adhesion to the epithelial cells. By treating bacteria with different mucins before infection, it is possible to study the ability of each mucin to inhibit the binding of the bacteria to the epithelial cells. In this chapter, we describe a protocol to culture MKN7 cells and *H. pylori* strain P12, treat *H. pylori* with mucin and co-culture MKN7 cells with mock-treated or mucin-treated *H. pylori* in vitro, and, finally, detect *H. pylori* bound to the MKN7 cells.

The protocol works well also with other cells, for example, from airways or intestine. An important consideration when choosing the eukaryote cell model (e.g., a cell line or primary cells) is that the cell model is firmly adherent to the 96-well cell culture plate. If it is not, some cell lines can adhere better to collagen-coated 96-well plates. The cells need to adhere so firmly to the 96-well plate that they do not wash off during the wash steps in the protocol. It is important to check that the cells still cover the plastic surface in all wells at the end of the assay, to ensure that artifacts are not introduced due to loss of cells. Furthermore, it is important that the cells are confluent before adding the bacteria, so that the bacteria adhere to the epithelial cells and not to the plastic surface of the 96-well cell culture plate.

The protocol works well with other bacterial species too. The current protocol is adjusted for *H. pylori*, which is slow growing. If working with fast-growing bacteria, overgrowth of the bacteria can be a problem. This issue can be solved by co-culturing the bacteria with the eukaryotic cells for 6–12 h instead of the 24 used in the current protocol and/or decreasing the number of bacteria added to the eukaryotic cells.

2 Materials

1. RPMI with UltraGlutamine I (Lonza).
2. Fetal bovine serum (FBS, Gibco).
3. Dulbecco's phosphate-buffered saline: without Ca^{++} and Mg^{++} (DPBS; Lonza).
4. Pen-Strep (5000 U Penicillin/mL, 5000 U Streptomycin/mL; Lonza).
5. Cell culture media: RPMI with UltraGlutamine I supplemented with 10% FBS (v/v) and 1% Pen-Strep (v/v).

6. Cell media (without antibiotics): RPMI with UltraGlutamine I supplemented with 10% FBS (v/v).
7. Cell culture flasks size 25 cm^2.
8. 96-well cell culture plates.
9. Trypsin/EDTA (170,000 U trypsin/L, 200 mg/L EDTA; Lonza).
10. Trypan Blue Solution: 0.4% in phosphate buffer saline (HyClone).
11. Brucella agar (Oxoid).
12. Defibrinated horse blood (Thermo Scientific).
13. Vancomycin hydrochloride from *Streptomyces orientalis* (Sigma).
14. Amphotericin B (BioWest).
15. Trimethoprim (Alfa Aesar).
16. Vitox mix (Oxoid).
17. Formaldehyde 4% aqueous solution.
18. Sterile phosphate-buffered saline (PBS; Medicago).
19. Rabbit anti-*H. pylori* serum.
20. HRP-conjugated donkey anti-rabbit IgG (Jackson).
21. Tween-20 (VWR).
22. Blocking reagent for ELISA (Roche).
23. Tetramethylbenzidine substrate (TMB, Sigma).
24. Sulfuric acid (VWR).

3 Methods

First, the gastric cell line MKN7 is grown up in sufficient numbers (upscale as needed, depending on the size of the planned experiment) from frozen stock, transferred to a 96-well plate, and grown until confluency. In parallel, *H. pylori* is grown up from frozen stock.

Then, the *H. pylori* strain P12 is pretreated with mucins (or mock pretreated, for the control) and added to the confluent MKN7 cells in a 96-well plate. After 24 h of co-culture, the plates are washed and the level *H. pylori* bound to the MKN7 cells investigated by an assay akin to enzyme-linked immunosorbent assay (ELISA) in the 96-well cell culture plate.

3.1 Cell Culture

For this assay, MKN7 cells are cultured at 37 °C in air supplemented with 5% CO_2, first in a 25 cm^2 for approximately 6 days and then for 72 h in a 96-well plate.

1. Take a stock vial of MKN7 cells from a −150 °C freezer.
2. Thaw the cells in a water bath at 37 °C.
3. Transfer the cells from the vial to a tube and add 5 mL of cell culture media. Carefully, pipette the mixture up and down.
4. Centrifuge at 1000×*g* for 5 min at room temperature (RT).
5. Remove the supernatant.
6. Resuspend the cells in 5 mL of cell culture media.
7. Transfer the mixture to a 25 cm^2 culture flask.
8. Incubate at 37 °C in air supplemented with 5% CO_2.
9. Grow the MKN7 cells for approximately 6 days and change the cell culture media every other day.
10. Check the confluency of the cells with a microscope. When the cells are 85–90% confluent in the flask, there are enough cells to proceed with the experiment following the layout in Fig. 1a (*see* **Note 1**).
11. Remove the cell culture media.
12. Wash with 10 mL of DPBS.
13. Aspirate the DPBS, and add 3 mL of trypsin/EDTA.
14. Incubate for 5 min at 37 °C.
15. In a microscope see if the cells have detached from the culture flask. If the cells are still adherent, incubate a couple of minutes longer.
16. Add 5 mL of cell culture media.
17. Transfer the cell suspension to a tube, and centrifuge at 1000×*g* for 5 min at RT.
18. Carefully remove the supernatant and add new cell culture media.
19. Determine the number of viable cells using the Trypan Blue exclusion technique or another suitable method.
20. Prepare a suspension of 4×10^5 cells/mL.
21. Seed 4×10^4 cells/well by adding 100 μL of the cell suspension from **step 20** in each well of the 96-well plate following the layout as exemplified in Fig. 1a, b.
22. After 48 h incubation (24 h before co-culture with *H. pylori*) at 37 °C in air supplemented with 5% CO_2, remove the cell culture media.
23. Wash the wells with 200 μL DPBS.
24. Add new cell media (without antibiotics, 100 μL/well).
25. When the cell layer is confluent, co-culture can start. The MKN7 cell layer is normally confluent 24 h after the change

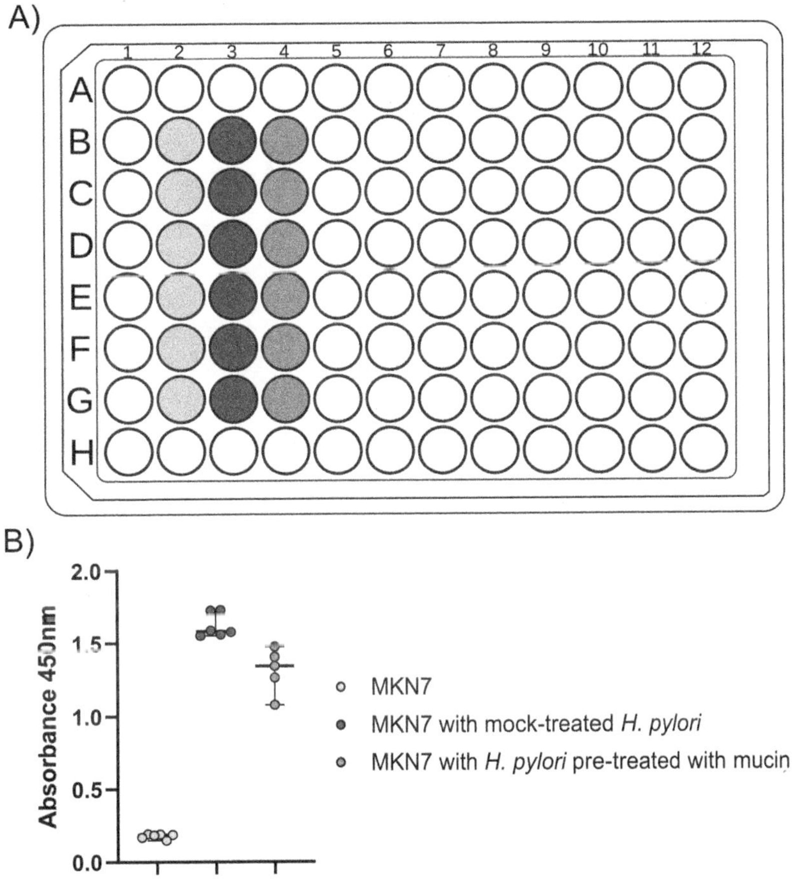

Fig. 1 Example of a 96-well cell culture plate layout and results of an assay measuring binding of mucin-treated and mock-treated *H. pylori* to MKN7 cells. (**a**) Example of an experiment layout in a 96-well cell culture plate. MKN7 cells (no *H. pylori* no mucin) in wells *2B* to *2G*; MKN7 cells co-cultured with mock-treated *H. pylori* in wells *3B* to *3G* and MKN7 cells co-cultured with *H. pylori* pretreated with mucin in wells *4B* to *4G*. (**b**) Result of a corresponding binding assay expressed in absorbance 450 nm. In light gray, no *H. pylori* were added to the cells therefore there is no binding. By pretreating *H. pylori* with mucins, less bacteria bound to the MKN7 cells (in gray) compared to mock-treated *H. pylori* (in dark gray)

of media to cell media (without antibiotics). Keep the 96-well cell culture plate at 37 °C in air supplemented with 5% CO_2 until *H. pylori* is added in method Subheading 3.4.

3.2 Bacteria Culture

For this assay, *H. pylori* strain P12 wild type is used.

1. Take out a stock vial of *H. pylori* P12 from the −80 °C freezer. Keep it in a benchtop cooler.
2. With a 10 μL loop, take *H. pylori* P12, and resuspend it in 50 μL of PBS. Mix by carefully pipetting up and down.
3. Transfer the 50 μL suspension of *H. pylori* P12 in PBS to Brucella agar plates supplemented with 10% defibrinated

horse blood (v/v), Vitox (10 mL for 500 mL agar), and antibiotic mix (Vancomycin 10 mg/L; Amphotericin B 4 mg/L; Trimethoprim 5 mg/L). Using a T-spreader, gently spread the suspension until the agar looks dry.

4. Grow *H. pylori* at 37 °C under microaerobic conditions for 4–5 days.
5. Re-streak the *H. pylori* onto new agar plate 2 days before the planned start of co-culture with MKN7 (*see* **Note 2**).

3.3 Pretreatment of H. pylori P12 with Mucins

1. Harvest *H. pylori* from the agar plates by adding 1 mL of PBS and gently scraping the bacteria from the agar surface with a T-spreader.
2. Aspirate *H. pylori* in PBS from the agar plate, and transfer it to a microcentrifuge tube.
3. Centrifuge at 412×*g* for 3 min.
4. Remove the supernatant and resuspend the *H. pylori* in cell media (without antibiotics).
5. Prepare a suspension of *H. pylori* $OD_{600} = 2.0$ in cell media (without antibiotics).
6. Dilute the mucin sample in cell media (without antibiotics) to a concentration of 250 μg/mL.
7. To treat *H. pylori* with mucin: add 70 μL of cell media (without antibiotics), 20 μL mucin diluted in **step 5**, and 10 μL of *H. pylori* suspension from **step 4** (volume per well).
8. Mock-treated *H. pylori*: add 90 μL of cell media (without antibiotics) and 10 μL of *H. pylori* suspension from **step 4** (volume per well).
9. Incubate *H. pylori* with mucin (**step 6**) and *H. pylori* without mucin (**step** 7) for 2 h at 37 °C in air supplemented with 5% CO_2.

3.4 Co-culture of MKN7 with H. pylori Treated with Mucins

The MKN7 cells are co-cultured according to the layout in Fig. 1 (*see* **Note 3**).

1. Aspirate the cell media from the cells seeded into the 96-well plate.
2. According to the layout in Fig. 1, add 100 μL of the following to the respective wells:
   2.1. Cell media (without antibiotics).
   2.2. Mock-treated *H. pylori* in cell media (without antibiotics).
   2.3. *H. pylori* pretreated with mucin in cell media (without antibiotics).
3. Incubate for 24 h at 37 °C in air supplemented with 5% CO_2.
4. Remove the media from the wells.

5. Wash the wells three times by gently adding the 300 μL PBS along the wall with a multichannel pipette or by using an automatic plate washer (*see* **Note 4**).
6. Fix the cells with cold buffered formaldehyde 4% aqueous solution (pH 7.0 ± 0.2) on ice for 10 min (100 μL/well). Check that the cell layers in the wells remain confluent/attached in a microscope.
7. If not proceeding immediately to the binding assay (method Subheading 3.5), add 100 μL 70% ethanol and keep at 4 °C overnight.

3.5 Determination of the Ability of Mucin to Inhibit H. pylori Binding to MKN7 Cells (ELISA Type Assay)

1. Aspirate the supernatant (4% formaldehyde or 70% ethanol, depending on if arriving from **step 6** or 7 from Subheading 3.4).
2. Gently wash the wells three times using a multichannel pipette or an automatic plate washer with 300 μL PBS containing 0.05% Tween (v/v).
3. Block unspecific binding sites with blocking reagent for ELISA (200 μL/well) for 1 h at RT.
4. Dilute the rabbit anti-*H. pylori* serum (primary antibody) 1: 1500 in blocking reagent for ELISA (*see* **Note 5**).
5. Remove blocking reagent for ELISA and add 100 μL of primary antibody (previous step) per well.
6. Incubate for 1 h at RT.
7. Dilute the secondary antibody, HRP-conjugated donkey anti-rabbit IgG, 1:10,000 in blocking reagent for ELISA.
8. Wash three times with 300 μL PBS containing 0.05% Tween (v/v), as per in **step 2**.
9. Add 100 μL of secondary antibody (**step** 7) per well.
10. Incubate for 1 h at RT.
11. Wash three times with 300 μL PBS containing 0.05% Tween (v/v), as per in **step 2**.
12. Add 100 μL of TMB substrate/well.
13. After 5 min, stop the reaction with 0.5 M sulfuric acid (100 μL/well).
14. Measure the absorbance at 450 nm in a suitable plate reader.

4 Notes

1. A 25 cm^2 flask is sufficient to grow enough cells to follow the layout on Fig. 1a. If more cells/wells are needed, then continue to grow the cells by expanding to flasks with a larger area.

2. One agar plate of *H. pylori* is sufficient to co-culture according to layout on Fig. 1a. If more bacteria are needed, re-streak into a higher number of agar plates.
3. A plate layout where technical replicates are arranged by column is used here to avoid possible errors in washing; in case there is an unnoticed defective channel in a plate washer that washes in columns (i.e., has a washing head with 8 or 16 channels). If the lab instead has a plate washer that washes in rows (i.e., has a washing head with 12 or 24 channels), put the replicates in a row-based layout instead.
4. If using an automatic plate washer, set up the washing program to slowly aspirate the supernatant and gently add the washing solution. This is to avoid the formation of aerosols and to avoid that the eukaryotic cells and/or bacteria detach. Additionally, because bacteria are alive, run an appropriated cleaning solution through the plate washer, and discard the biological waste appropriately.
5. Other types of commercially available anti-*Helicobacter* antibodies can also be used.

Acknowledgements

This work was supported by the Swedish Research Council (2019-01152) and the Swedish Cancer Association (23 2868 Pj).

References

1. McGuckin MA, Linden SK, Sutton P, Florin TH (2011) Mucin dynamics and enteric pathogens. Nat Rev Microbiol 9(4):265–278
2. Linden SK, Sheng YH, Every AL, Miles KM, Skoog EC, Florin TH et al (2009) MUC1 limits Helicobacter pylori infection both by steric hindrance and by acting as a releasable decoy. PLoS Pathog 5(10):e1000617
3. Quintana-Hayashi MP, Navabi N, Mahu M, Venkatakrishnan V, Fernandez HR, Haesebrouck F et al (2017) Neutrophil elastase and interleukin 17 expressed in the pig colon during Brachyspira hyodysenteriae infection synergistically with the pathogen induce increased mucus transport speed and production via mitogen-activated protein kinase 3. Infect Immun 85(8):e00262-17
4. Linden SK, Sutton P, Karlsson NG, Korolik V, McGuckin MA (2008) Mucins in the mucosal barrier to infection. Mucosal Immunol 1(3): 183–197
5. Lindén SK, Wickström C, Lindell G, Gilshenan K, Carlstedt I (2008) Four modes of adhesion are used during helicobacter pylori binding to human mucins in the oral and gastric niches. Helicobacter 13(2):81–93
6. Padra JT, Sundh H, Jin C, Karlsson NG, Sundell K, Linden SK (2014) Aeromonas salmonicida binds differentially to mucins isolated from skin and intestinal regions of Atlantic salmon in an N-acetylneuraminic acid-dependent manner. Infect Immun 82(12): 5235–5245
7. Skoog EC, Padra M, Aberg A, Gideonsson P, Obi I, Quintana-Hayashi MP et al (2017) BabA dependent binding of Helicobacter pylori to human gastric mucins cause aggregation that inhibits proliferation and is regulated via ArsS. Sci Rep 7:40656
8. Lindén S, Nordman H, Hedenbro J, Hurtig M, Borén T, Carlstedt I (2002) Strain- and blood group-dependent binding of

Helicobacter pylori to human gastric MUC5AC glycoforms. Gastroenterology 123(6):1923–1930

9. Padra JT, Loibman SO, Thorell K, Sundh H, Sundell K, Lindén SK (2022) Atlantic salmon mucins inhibit LuxS-dependent A. salmonicida AI-2 quorum sensing in an N-acetylneuraminic acid-dependent manner. Int J Mol Sci 23(8): 4326
10. Linden S, Mahdavi J, Semino-Mora C, Olsen C, Carlstedt I, Boren T, Dubois A (2008) Role of ABO secretor status in mucosal innate immunity and H. pylori infection. PLoS Pathog 4(1):e2
11. Muthupalani S, Ge Z, Joy J, Feng Y, Dobey C, Cho HY et al (2019) Muc5ac null mice are predisposed to spontaneous gastric antro-pyloric hyperplasia and adenomas coupled with attenuated H. pylori-induced corpus mucous metaplasia. Lab Investig 99(12): 1887–1905
12. McGuckin MA, Every AL, Skene CD, Linden SK, Chionh YT, Swierczak A et al (2007) Muc1 mucin limits both Helicobacter pylori colonization of the murine gastric mucosa and associated gastritis. Gastroenterology 133(4): 1210–1218

Chapter 5

Profiling of Anti-glycan IgA and IgG in Oral Fluid Using a Glycan Microarray

Nadezhda V. Shilova, Svetlana M. Polyakova, Marina V. Lavrenteva, and Nicolai V. Bovin

Abstract

Secretory immunoglobulins (sIg), primarily IgA, control the microflora of the mucous membrane, protecting it from pathogens, as well as promoting its colonization by commensals. Knowledge of the specificity of sIgA (as well as sIgG) is necessary to understand their functions. We present here a protocol for using printed glycan array to profile the glycan specificity of sIg in oral fluid, with particular attention to sample preparation of this difficult-to-analyze material. The list of glycans in the array includes oligosaccharides typical of mucin *O*-chains, as well as bacterial polysaccharides, which are another full-fledged "player" in the sIg–mucin–microbiota system.

Key words Mucosal immunity, Oral fluid, Glycan microarray, PGA, Secretory IgA, Mucins, Anti-glycan antibodies, Microbiota

1 Introduction

Secretory immunoglobulins play important role in local communications between microbial community, host mucin interface, and immune system. The main classes of immunoglobulins in oral fluid are IgA and IgG (named further as sIgA and sIgG in order to differentiate them from blood antibodies). It's believed that normally sIgG enters the oral cavity from the peripheral blood circulation by passive transmucosal diffusion [1], although some of it is produced locally in the salivary glands or gingiva [2]. Its level significantly increased during infections and other diseases, as well as vaccination [3–5]. Once contacted a pathogen, sIgG eliminates it via complement activation and phagocytosis enhancement, i.e., plays a protective role [6–8].

sIgA is a dimeric molecule (higher oligomers also exist) fastened with joining (J) chain and decorated with secretory component (SC) [9]. sIgA has several functions. First of all these are

Inka Brockhausen (ed.), *Dynamics of Bacteria-Mucus Interactions*, Methods in Molecular Biology, vol. 2942,
https://doi.org/10.1007/978-1-0716-4627-4_5,

neutralization of pathogenic microorganisms (viruses, bacteria, fungi) by the opsonization, blocking of attachment to the cell surface receptors, inhibition of pathogen mobility and their growth as well as preventing the access of antigen to immune system and implemented through the canonical interaction *antigen*-F_{ab} [8–10]. Noncanonical interactions occur through the binding of highly glycosylated SC with microbial lectins [11, 12]. IgA does not activate the complement system (as some IgG isotypes does) and even inhibits it [13], although activation of the lectin pathway by recruiting mannose-binding lectin has been described, when sIgA and lectin act synergistically [14].

sIgA promotes the colonization of the oral cavity by commensal microbiota using the same mechanisms as when interacting with pathogenic microorganisms, i.e., through F_{ab} and SC, but without the manifestation of cytotoxicity (noninflammatory pathway), maintaining the balance between the host and the microflora. These properties are assumed to be characteristic of naturally occurring sIgA, the antigen specificity of which is evolutionarily selected [15–17]. sIgA is especially effective in implementing the so-called "immune exclusion" of antigens and toxins when it forms a complex with mucin through cysteine residues [18], although a mechanism involving SC is also considered [19–21].

A significant proportion of microbial antigens are the glycans (oligo- and polysaccharides), available in isolated or synthetic form, and which are convenient for immobilization. Arrays of these glycans as well as glycans of mammalian origin (often in format of printed glycan array, PGA) are used in studies of glycan-binding proteins (CBPs), including human blood anti-glycan antibodies [22–25]. Oral fluid is one of the most difficult objects for studying immunoglobulins due to its viscosity, the presence of mucins of different nature and antigenicity, non-Ig CBP, enzymes, etc. To overcome the associated problems when analyzing secretory immunoglobulins, alternate methods are being sought, in particular, IgA is isolated from the pooled saliva samples or from the culture fluid of the mucosa-derived B cells [21, 26, 27]. Complete as possible knowledge about the carbohydrate targets of secretory Ig is necessary for, firstly, to study the balance between microorganisms, epithelium, secretory mucins and immunoglobulins, and secondly, to elucidate the reasons for changes in the reactivity of the mucous membrane when exposed to a pathogen or in autoimmune disorders. This knowledge is certainly necessary for the development of methods for diagnosing diseases caused by pathogens (or other microbes) and monitoring the effectiveness of nasal (and other) vaccines [28].

Here we propose a sample preparation protocol that includes [1] high-speed centrifugation and [2] filtration of the diluted supernatant through a 0.22 μm filter, which together removes most of the mucins and bacteria, and [3] lowering the incubation temperature to inhibit bacterial enzyme activity. These measures result in a low background when determining the glycan binding profile of sIgG and sIgA in oral fluid.

2 Materials

Prepare all solutions in the bidistilled water and using the analytical grade reagents unless indicated otherwise. Prepare and store all reagents at +4 °C unless indicated otherwise. Perform all experiments with human fluids wearing laboratory gloves and glasses. Follow all local waste disposal regulations when disposing of waste materials.

2.1 Oral Fluid Sample Preparation

1. Frozen human oral fluid in 0.5 mL aliquots in 1.5 mL microcentrifuge tubes (whole oral fluid collected in the morning 1–2 h before breakfast and teeth cleaning after rinsing the mouth with a tap water). Store at −70 °C.
2. 0.22 μm, 13 mm polyvinylidene fluoride (PVDF) syringe filters (*see* **Note 1**).
3. Microtube centrifuge.
4. −70 °C freezer for storage of oral fluids samples.
5. 10 mM phosphate-buffered saline (PBS): dissolve 0.35 g NaH_2PO_4, 1.28 g Na_2HPO_4, and 8 g NaCl in approximately 900 mL of water, and then make up to 1 L with water. Check that the pH of the resulting solution is 7.4.

2.2 Microarray Binding Assays

1. Glycan microarrays (Array-700, GlycoNZ, Auckland, New Zealand) with barcodes, containing 480 oligo- and 295 polysaccharides (or similar microarray). The ligands description done in the Table 1. Store at +20 °C under vacuum.
2. Block buffer: dissolve 3.8 g of boric acid in 400 mL of water, add 1000 μL of ethanolamine and 800 μL of the Tween 20, mix well, and check that the pH of the resulting solution is 8.5 (*see* **Note 2**).
3. Sample buffer (PBS-0.1%T-1%BSA): 0.1% (v/v) Tween 20 and 1% of the bovine serum albumin, BSA, fraction V, in the PBS.
4. Washing buffer (PBS-0.05T): 0.05% (v/v) Tween 20 in PBS.
5. Paper towel or Kimwipes.
6. AlexaFluor647-labeled goat anti-human IgG, Fcγ-specific (for example, Jackson ImmunoResearch, West Grove, PA, USA), 1.5 mg/mL solution in PBS, 0.05% sodium azide, and 15 mg/mL BSA (as provided by the manufacturer).
7. Cy3-labeled goat anti-human IgA, α-chain specific (for example, Jackson ImmunoResearch, West Grove, PA, USA),

Table 1
The main groups of glycoligands in composition of microarray

Designation, systematic, or trivial name	
1. Mucin-type (core and terminal fragments) and related structures	2. Sialoglycans
3. Blood groups related structures (ABO, Lewis, P, Ii, etc.)	4. *N*-acetyllactosamine containing glycans (including glycoprotein *N*-chains)
5. Glucuronic acid (GlcA), GlcA-containing disaccharides, hyaluronic acids	6. Sulfated glycans
7. Mannose-containing oligosaccharides	8. Lactose derivatives
9. Tumor-associated carbohydrate antigens	10. Glycosphingolipids
11. Saccharides—fragments of bacterial and plant polysaccharides	12. Bacterial capsular and *O*-polysaccharides

1.5 mg/mL solution in PBS, 0.05% sodium azide, and 15 mg/mL BSA (as provided by the manufacturer).

8. Five-pcs microscope slide boxes.
9. +4 °C refrigerator.
10. Orbital shaker.
11. Orbital thermoshaker.
12. Vortex.
13. Mini-centrifuge for slides.
14. Microarray fluorescence scanner (for example, InnoScan 1100 AL, Innopsys, Carbonne, France, or another microarray scanner).
15. Incubation chamber with a lid.

2.3 Software

1. Fluorescence scanner software (Mapix 8.1.1 if Innoscan 1100AL is used or another software connected with a scanner).
2. TIFF image analysis software (for example, Scan Array Express 4.0, PerkinElmer, Waltham, MA, USA).
3. Microsoft Excel.

3 Methods

3.1 Oral Fluid Sample Preparation

1. Remove oral fluid aliquots from the −70 °C freezer and thaw on ice. We typically process up to eight samples at a time.
2. Once thawed centrifuge the sample (10,500×*g*, 5 min), aspirate 100 μL, and dilute up to 800 μL with the sample buffer. Mix well using vortex, and then centrifuge the diluted sample again (10,500×*g*, 5 min).

3. Take 800 μL of supernatant, filter it through the 0.22 μm, and make up to 1 mL with the sample buffer.

3.2 Glycan-Binding Profiling of Mucosal IgG and IgA

3.2.1 Microarray Assay

1. Put microarrays to the Couplin Jar (or plastic box for five-pcs microscope slides). Pour enough blocking buffer just to cover the full printed area and gently shake on orbital shaker (100 rpm) for 2 h (*see* **Note 3**).
2. Remove microarrays from the jar, and wash them dipping into the glass with the washing buffer 10 times. Collect excess buffer by touching corners of arrays to a paper towel.
3. Put the microarrays to the incubation chamber on a back side, so that the barcode is facing up and reads correctly (pour water on a bottom of the chamber for humidifying).
4. Add 900 μL of the sample on the array (one sample—one array) carefully spreading it to the surface using automatic pipette (*see* **Note 4**).
5. Close the lid of the chamber and incubate microarrays at +4 °C in a fridge under the slight shake (30 rpm) overnight.
6. Remove samples from the microarrays, and wash them dipping consequently into the two glasses with the washing buffer 10 times in each. Collect excess buffer by touching corners of arrays to a paper towel, and dry off the backside of the slide (*see* **Note 5**).
7. Dilute both the anti-IgG-Alexa647 and anti-IgA-Cy3 antibodies in the sample buffer to a final concentration of 1 μg/mL in one tube (*see* **Note 6**).
8. Put the microarrays to the incubation chamber, and add 900 μL of solutions containing both anti-IgG and anti-IgA antibodies.
9. Close the lid of the chamber, and incubate microarrays at +37 ° C in the thermoshaker under the slight shake for 1 h.
10. Remove solution from the microarrays, and wash them dipping consequently into the two glasses with the washing buffer 10 times in each.
11. Finally wash microarrays in water, collect excess buffer by touching corners of arrays to a paper towel, and spin dry in mini-centrifuge.
12. Place microarrays to a five-pcs microscope slide box and keep in the dark until scanning (*see* **Note 7**).
13. Scanning procedure has been described in detail in several protocols such as [29] or [30]. In brief, scan microarrays using a microarray fluorescence scanner using two types of laser, 532 and 635 nm, for detection of IgA and IgG, correspondingly (*see* **Note 8**). In our laboratory we have Innoscan

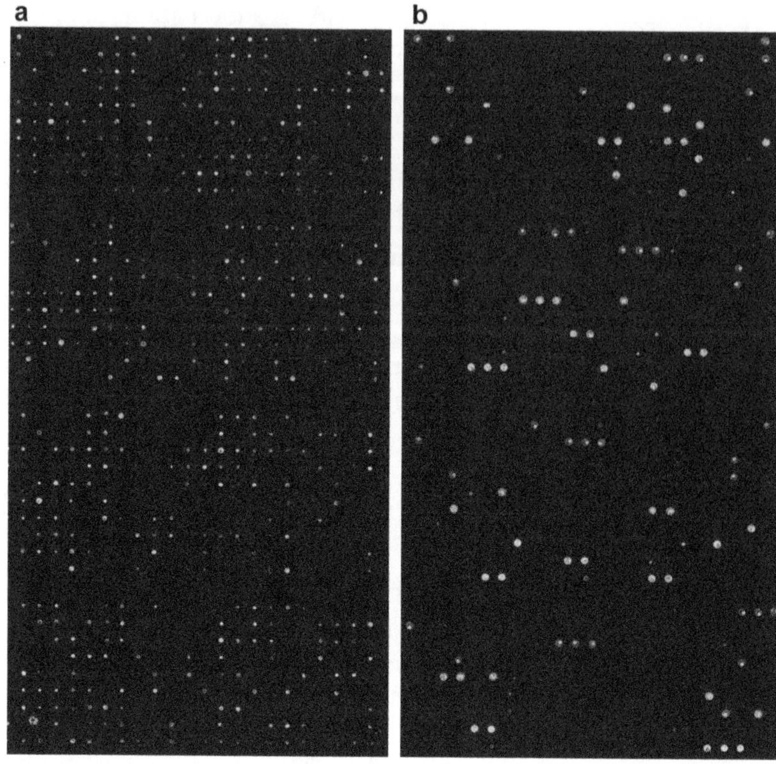

Fig. 1 Interaction of sIgA (**a**) and sIgG (**b**) from the oral fluid of a healthy donor with glycan array. sIgG visualized with Alexa647-labeled goat anti-human IgG and sIgA with Cy3-labeled goat anti human IgA. Images are presented in a rainbow color scale

1100AL, allowing to perform scanning on both wavelength simultaneously using Mapix software. Recommended scanning resolution is 10 μm. The examples of obtained images are seen in Fig. 1a, b.

3.2.2 Data Extraction and Analysis

Data extraction has been also described in detail in, e.g. [30]. In brief,

1. After scanning transfer obtained TIFF images to an image processing program. In our laboratory we use ScanArrayExpress 4.0. Other programs can also be used.
2. Using the GAL file (file containing information about the ligand identification numbers as well as their positions) convert image to the csv file (raw file) with digital data from the array.
3. Convert raw file to Excel xlsx file using any appropriate approach, for example, Excel macro. Obtained files from all tested samples can be collected in one file, sorted and graphed (*see* **Note 9**).

4 Notes

1. Other 0.22 μm filters with low protein binding can be used.
2. Use ethanolamine or boric acid to adjust pH in the Block buffer. Tween 20 is very viscose liquid; wait 2–3 s after aspiration and dispensing.
3. To avoid microdroplets of saliva released during breathing from contaminating the chip, we recommend working in a COVID-type mask.
4. Do not touch microarray surface by a pipette tip.
5. Do not allow microarray drying between incubation stages.
6. All needed solutions (except buffers) have to be freshly prepared (immediately before analysis).
7. After the arrays have been dried after the last incubation, they are ready to be immediately scanned. We recommend to scan right after the experiment or at least next day (stored at room temperature in the dark).
8. Scan settings are dependent on the sample and fluorescent label used. The microarray fluorescence scanner has a maximum relative fluorescence unit (RFU) of 65,535 for a saturated spot. It is best to have the strongest signals below this saturation for the most samples varying the PMT and/or laser power. Use the preview option to prevent having saturated signals.
9. Using the proposed protocol, an investigator takes into account the immunoglobulins in the liquid part of an oral fluid, which are not bound to mucins or bacteria removed during the sample preparation. It should be also noted that the total sIgA loss is less than 5%, in case of sIgG—about 10 % (unpublished data).

Acknowledgements

The authors express gratitude for the preparing of the printed glycan arrays to A. Nokel (Semiotik LLC, Moscow, Russia).

References

1. Kim K, Keller MA, Heiner DC (1992) Immunoglobulin G subclasses in human colostrum, milk and saliva. Acta Paediatr 81:113–118
2. Brandtzaeg P (2007) Do salivary antibodies reliably reflect both mucosal and systemic immunity? Ann N Y Acad Sci 1098:288–311
3. Tsunetsugu-Yokota Y, Ito S, Adachi Y et al (2022) Saliva as a useful tool for evaluating upper mucosal antibody response to influenza. PLoS One 17:e0263419
4. Gürsoy M, Rautava J, Pussinen P et al (2023) Salivary IgA and IgG antibody responses against periodontitis-associated bacteria in Crohn's disease. Int J Mol Sci 24:2385
5. Lahdentausta L, Kivimäki A, Oksanen L et al (2022) Blood and saliva SARS-CoV-2 antibody

levels in self-collected dried spot samples. Med Microbiol Immunol 211:173–183

6. Anderson DM, Ebersole JL, Novak MJ (1995) Functional properties of nonhuman primate antibody to Porphyromonas gingivalis. Infect Immun 63:3245–3252
7. McArthur WP, Clark WB (1993) Specific antibodies and their potential role in periodontal diseases. J Periodontol 64:807–818
8. Challacombe SJ, Shirlaw PJ (2005) Immunology of diseases of the oral cavity. In: Mestecky J, Strober W, etc. (eds) Mucosal immunology, 4th edn. Elsevier, Amsterdam, pp 1517–1546
9. Hockenberry A, Slack E, Stadtmueller BM (2023) License to clump: secretory IgA structure–function relationships across scales. Ann Rev Microbiol 77:645–668
10. Li Y, Jin L, Chen T (2020) The effects of secretory IgA in the mucosal immune system. Biomed Res Int 2020:1–6
11. Kaetzel CS (2005) The polymeric immunoglobulin receptor: bridging innate and adaptive immune responses at mucosal surfaces. Immunol Rev 206:83–99
12. Oliveira IR, de Araújo AN, Bao SN, Giugliano LG (2001) Binding of lactoferrin and free secretory component to enterotoxigenic Escherichia coli. FEMS Microbiol Lett 203: 29–33
13. Russell MW, Mansa B, Reinholdt J, Kilian M (1990) Inhibition of complement activation by human IgA antibodies and their Fab fragments. In: MacDonald T, Challacombe S, etc. (eds) Advances in mucosal immunology. Springer, Dordrecht, pp 869–873
14. Roos A, Bouwman LH, van Gijlswijk-Janssen DJ et al (2001) Human IgA activates the complement system via the mannan-binding lectin pathway. J Immunol 167:2861–2868
15. Slack E, Balmer ML, Fritz JH, Hapfelmeier S (2012) Functional flexibility of intestinal IgA—broadening the fine line. Front Immunol 3:100
16. Stoel M, Jiang H-Q, van Diemen CC et al (2005) Restricted IgA repertoire in both B-1 and B-2 cell-derived gut plasmablasts. J Immunol 174:1046–1054
17. de la Cruz Peña MJ, Gonzalez-Granado LI, Garcia-Heredia I et al (2021) Minimal-moderate variation of human oral virome and microbiome in IgA deficiency. Sci Rep 11: 14913
18. Biesbrock AR, Reddy MS, Levine MJ (1991) Interaction of a salivary mucin-secretory immunoglobulin A complex with mucosal pathogens. Infect Immun 59:3492–3497
19. Phalipon A, Cardona A, Kraehenbuhl J-P et al (2002) Secretory component. Immunity 17: 107–115
20. Pabst O, Slack E (2020) IgA and the intestinal microbiota: the importance of being specific. Mucosal Immunol 13:12–21
21. Sterlin D, Fadlallah J, Adams O et al (2020) Human IgA binds a diverse array of commensal bacteria. J Exp Med 217:e20181635
22. Huflejt ME, Vuskovic M, Vasiliu D et al (2009) Anti-carbohydrate antibodies of normal sera: findings, surprises and challenges. Mol Immunol 46:3037–3049
23. Oyelaran O, McShane LM, Dodd L, Gildersleeve JC (2009) Profiling human serum antibodies with a carbohydrate antigen microarray. J Proteome Res 8:4301–4310
24. Stowell SR, Arthur CM, McBride R et al (2014) Microbial glycan microarrays define key features of host-microbial interactions. Nat Chem Biol 10:470–476
25. Luetscher RND, McKitrick TR, Gao C et al (2020) Unique repertoire of anti-carbohydrate antibodies in individual human serum. Sci Rep 10:15436
26. Schneider C, Illi M, Lötscher M et al (2017) Isolation of antibodies from human plasma, saliva, breast milk, and gastrointestinal fluid. In: Kaveri S, Bayry O (eds) Natural antibodies, Methods in molecular biology. Springer, Clifton, pp 23–31
27. Sharanova NE, Ninnemann J, Bondareva MA et al (2017) Analysis of the specificity of IgA antibodies produced in the mouse small intestine. Mol Biol 51:813–818
28. Lavelle EC, Ward RW (2022) Mucosal vaccines—fortifying the frontiers. Nat Rev Immunol 22:236–250
29. Busch J, McBride R, Head SR (2010) Production and application of glycan microarrays. In: Chittur S (ed) Microarray methods for drug discovery, Methods in molecular biology. Springer, Clifton, pp 269–282
30. Temme JS, Gildersleeve JC (2022) General strategies for glycan microarray data processing and analysis. In: Kilcoyne M, Gerlach O (eds) Glycan microarray, Methods in molecular biology. Springer, Clifton, pp 67–87

Chapter 6

Assays for Anti-*Helicobacter pylori* Effects of Lactic Acid Bacteria Secretions

Jean W. Marbou Takougoum, Samantha A. Whiteside, Daelyn G. Carroll, Mahi M. Mohiuddin, Stéphane Bronner, Sylvie Binda, and Carole Creuzenet

Abstract

Helicobacter pylori (HP) is a chronic bacterial pathogen that can cause gastric ulcers and cancers. The treatment for HP infections relies on a combination of antibiotics and proton pump inhibitors. Due to rising antibiotic resistance, curtailing HP-induced diseases requires identifying new anti-HP molecules that can reduce HP viability or impair its virulence, including its ability to cause inflammation. Oral intake of lactic acid bacteria (LAB) was shown to enhance the treatment of HP infection, and we demonstrated that LAB secretions can affect HP's growth and virulence factor production. This article describes a panel of in vitro assays that can be used to screen the anti-HP properties of multiple LAB secretions for the identification of novel anti-HP agents. The assays pertain to effects on viability, motility, urease production, and ability to elicit inflammation in gastric cells.

Key words *Helicobacter pylori*, Lactic acid bacteria, Urease, Inflammation, Cytokines

1 Introduction

The bacterium *Helicobacter pylori* (HP) colonizes about 50% of the world population chronically, leading to gastric ulcers or cancers in 2–10% of infections [1]. The standard treatment for HP infections relies on combinations of two or three antibiotics, among clarithromycin, metronidazole, amoxicillin, tetracycline, levofloxacin, and furazolidone, in addition to proton pump inhibitors and bismuth [2–5]. However, resistance to all these antibiotics is rising dramatically, reaching 60% for metronidazole, 38% for levofloxacin, 31% for clarithromycin, and 12% for clarithromycin and metronidazole combined in the USA and/or worldwide [6, 7]. Resistance to amoxicillin (2.6%), tetracycline (0.87%), and furazolidone (1.3%) is also observed though at a lower extent. Due to rising antibiotic resistance, curtailing HP-induced diseases requires identifying new

Inka Brockhausen (ed.), *Dynamics of Bacteria-Mucus Interactions*, Methods in Molecular Biology, vol. 2942,
https://doi.org/10.1007/978-1-0716-4627-4_6,

anti-HP agents that will reduce HP viability or impair its virulence, including its ability to cause inflammation. This is in line with the WHO classification of HP as high priority pathogen for research into alternative antimicrobials since 2017 [8]. Multiple studies in both mice and humans have suggested that ingestion of lactic acid bacteria (LAB) can enhance the treatment of HP infection [9–12]. Specifically, LAB have been shown to prevent HP colonization, gastric tissue lesions and gastritis [13, 14] or helped clear HP infections [15–17]. Our team and others have shown that in vitro exposure to LAB reduced HP growth and its production of the essential virulence factor urease [18, 19] and that pretreatment with LAB also decreased HP adherence to gastric cells and HP-induced secretion of pro-inflammatory cytokines [18, 20, 21]. Specifically, we have identified 18 LAB strains out of 25 strains tested that have anti-HP effects as they decrease HP viability and decrease HP's production of urease, its motility, and/or its ability to elicit pro-inflammatory IL-8 from gastric cells [18]. Collectively, these studies strongly suggest that LAB produce bioactive molecules that can impact HP growth and virulence factors. Most studies investigate only a few LAB strains and a few phenotypes of a single HP strain at a time. We designed an in vitro framework that has so far allowed us to screen 25 LAB strains for 7 phenotypes against 2 different HP strains [18]. In this manuscript, we describe in vitro assays to study the anti-HP properties of LAB secretions regarding viability, motility, urease production, and interactions with gastric cells (*see* **Note 1**).

2 Materials and Equipment

2.1 Bacterial Strains and Human Cells

1. *H. pylori* strains NCTC 11637 (ATCC 43504) and SS1 as freezer stocks in 25% glycerol prepared by adding 1 volume of 50% glycerol to 1 volume of HP suspension adjusted at OD_{600} 1.5 and stored at -80 °C.
2. Lactic acid bacteria strains as freezer stocks (−80 °C) in 25% glycerol prepared by adding 1 volume of 50% glycerol to 1 volume of LAB cultured in de Man–Rogosa–Sharpe(MRS) broth and adjusted at OD_{600} of 1.5. Individually freeze-dried bacteria can be used to generate freezer stocks after reconstitution in MRS (10 mg of powder in 5 mL media) and growth for 24 h in anaerobic conditions without agitation, followed by subculture (dilution 1/25) in MRS media in the same conditions. Mixed freeze-dried bacteria are inoculated from powder (10 mg of powder in 5 mL MRS media) for each experiment (*see* **Note 2**).
3. Frozen human gastric adenocarcinoma cell AGS (ATCC CRL-1739) stocks in 5% DMSO prepared by adding 50 μL of DMSO in 950 μL of AGS cells at 2×10^5 cells/mL and stored at −80 °C.

2.2 Growth Media and Supplements

1. de Man–Rogosa–Sharpe (MRS) broth and agar (Becton, Dickinson and Company): Proteose peptone No. 3 10 g/L, beef extract 10 g/L, yeast extract 5 g/L, glucose 20 g/L, polysorbate 80 1 g/L, ammonium citrate 2 g/L, sodium acetate 5 g/L, magnesium sulfate 0.1 g/L, manganese sulfate 0.05 g/L, dipotassium phosphate 2 g/L, and agar 15 g/L (as needed).
2. Modified de Man–Rogosa–Sharpe (mMRS) broth: Proteose peptone No. 3 10 g/L, yeast extract 5 g/L, glucose 20 g/L, ammonium citrate 2 g/L, sodium acetate 5 g/L, magnesium sulfate heptahydrate 48 g/L, and manganese (II) sulfate monohydrate 51 g/L. All reagents combined in 1 L ddH_2O are autoclaved and allowed to cool down before addition of the following compounds at the indicated final concentrations: 11 mM dipotassium phosphate (2 g/L) and 0.12 mM iron (II) sulfate heptahydrate (33 g/L) (*see* **Note 3**).
3. Supplemented Columbia blood agar (CBA) plates: These comprise Columbia blood agar base (Becton, Dickinson and Company) made of pancreatic digest of casein 10 g/L, proteose peptone No. 35 g/L, yeast extract 5 g/L, beef heart infusion 3 g/L, corn starch 1 g/L, sodium chloride 5 g/L, and agar 15 g/L. All reagents combined in 1 L ddH_2O are autoclaved and allowed to cool down before addition of the following supplements at the indicated final concentrations: FBS (10%), sodium pyruvate (50 μg/mL), trimethoprim (5 μg/mL), vancomycin (10 μg/mL), and amphotericin B (5 μg/mL) (*see* **Note 4**).
4. Supplemented Brain Heart Infusion-Yeast Extract Broth (BHI): These comprise brain heart infusion 6.0 g/L, peptic digest of animal tissue 6.0 g/L, sodium chloride 5.0 g/L, glucose 3.0 g/L, pancreatic digest of gelatin 14.5 g/L, disodium phosphate 2.5 g/L, and yeast extract 1.5 g/L. All reagents combined in 1 L ddH_2O are autoclaved and allowed to cool down before addition of the same supplements as in CBA plates above.
5. Trimethoprim (Sigma, Quality level 200, purity ≥98.5%).
6. Vancomycin (Sigma, pharmaceutical primary standard).
7. Amphotericin B (Sigma, pharmaceutical secondary standard, quality level 300).
8. Sodium pyruvate (Thermo Fisher, purity 99%).
9. 0.85% saline: Sodium chloride 8.5 g/L, 1 L ddH_2O.
10. Glycerol (Thermo Fisher).
11. Anaerobic gas packs (Becton, Dickinson and Company).
12. GasPak EZ Campy sachets (Becton, Dickinson and Company).

2.3 Tissue Culture and Cytokine Assay Reagents

1. F-12 K Nutrient Mixture (Kaighn's Modification) (Wisent).
2. Fetal bovine serum (FBS) heat inactivated (Thermo Fisher).
3. 1x phosphate-buffered saline (1xPBS): 137 mM NaCl, 2.7 mM KCl, 10 mM Na_2HPO_4, 1.8 mM KH_2PO_4, pH 7.4.
4. 0.25% Trypsin-EDTA (Thermo Fisher).
5. Gibco Minimum Essential Media (MEM) Non-Essential Amino Acids (NEAA) (Thermo Fisher).
6. 0.4% Trypan blue.
7. Human IL-8 ELISA kit (eBioscience).

2.4 Reagents for Urease Assay

1. Buffered hypochlorite reagent: Prepare 100 mL of reagent in ddH_2O using 1 g sodium phosphate and 4.35 mL of sodium hypochlorite solution (sold as 10%–15% solution, pH 13, Sigma). Make this solution the day of each experiment (*see* **Note 5** *for safety*).
2. 10 mM sodium phosphate buffer pH 7: To prepare 1 L, mix 5.8 mL of 1 M Na_2HPO_4 with 4.3 mL of 1 M NaH_2PO_4.
3. 1 M urea in sodium phosphate buffer, pH 7: Prepare 100 mL at 60 g/L. Make this solution the day of each experiment.
4. Phenylphenol-nitroprusside reagent (PPN): Prepare 100 mL at 32.22 g/L of 2-phenylphenol Na•$4H_2O$ salt and 0.15 g/L sodium nitroprusside. Prepare weekly, stored at 4 °C, protected from light (*see* **Note 5**).
5. 5% citrate reagent (pH 7): Prepare 100 mL at 50 g/L and adjust pH to 7 with HCL.
6. 1 M urea solution: Prepare 10 mL at 60 g/L in ddH_2O. Make this solution the day of each experiment.
7. Urease reaction buffer (pH 7): Prepare daily a volume of 100 mL by mixing 50 μL of 1 M urea in 10 mM sodium phosphate buffer (made at **step 3** above) with 50 mL citrate reagent (made at **step 5** above) and 50 mL sodium phosphate buffer (made at **step 2** above).
8. Ammonium sulfate-concentrated standard solution (100 mM): Prepare 100 mL at 13.21 g/L. Store at 4 °C for up to 2 weeks. Working standards solutions are prepared daily (*see* **step 3,** Subheading 3.6.2).

2.5 Plasticware

1. 15 × 100 mm Petri dishes.
2. Square Petri dish with grid (36 13 × 13 mm squares) (Thermo Fisher).
3. 96-well tissue culture plates.
4. T25 flask-vented lid (Thermo Fisher).
5. T75 flask-vented lid (Thermo Fisher).

2.6 Equipment

1. 12-channel 5–50 μL adjustable-volume micropipette.
2. 12-channel 50–200 μL adjustable-volume micropipette.
3. Serological pipettes (5 mL and 10 mL).
4. Anaerobic jars.
5. Tri-gas incubator generating a 5% O_2 and 10% CO_2 microaerobic atmosphere.
6. Shaker incubator.
7. CO_2 incubator (5% CO_2).
8. Spectrophotometer.
9. Microplate reader.
10. Water bath set at 37 °C.
11. Biosafety cabinet (BSC).
12. pH meter and pH indicator strips.
13. Hemocytometer.
14. Centrifuge.
15. Autoclave.
16. Fume hood.

3 Methods

All the methods are described with volumes/dilutions/concentrations that allow preparing one 96-well exposure plate to screen four LAB strains/HP strain, with and without urea, along with LA, LN, and BHI controls. The BHI control only comprises optimal BHI growth media for HP while the LA and LN controls comprise LAB growth media first acidified with lactic acid to mimic LAB growth and then used as such (LA) or after neutralization (LN) (*see* **Note 6**). The specifics about their preparation are described in **step 5** below. A schematic diagram of all the assays described is reported in Fig. 1.

3.1 LAB Growth

Although only 400 μL of pH-adjusted LAB supernatant is needed per LAB strain and per HP strain exposed, all LAB cultures are set up at a 25 mL scale to ensure growth consistency and for practicality of pH adjustments and filter sterilization. The process is delineated in Fig. 1.

1. This section requires four MRS plates. Streak four LAB strain freezer stocks on one MRS agar plate each and incubate for 24 h at 37 °C in anaerobic conditions, using an anaerobic gas pack and airtight canister.
2. For each LAB strain, take three colonies, resuspend together in 5 mL MRS broth in loosely capped culture tubes

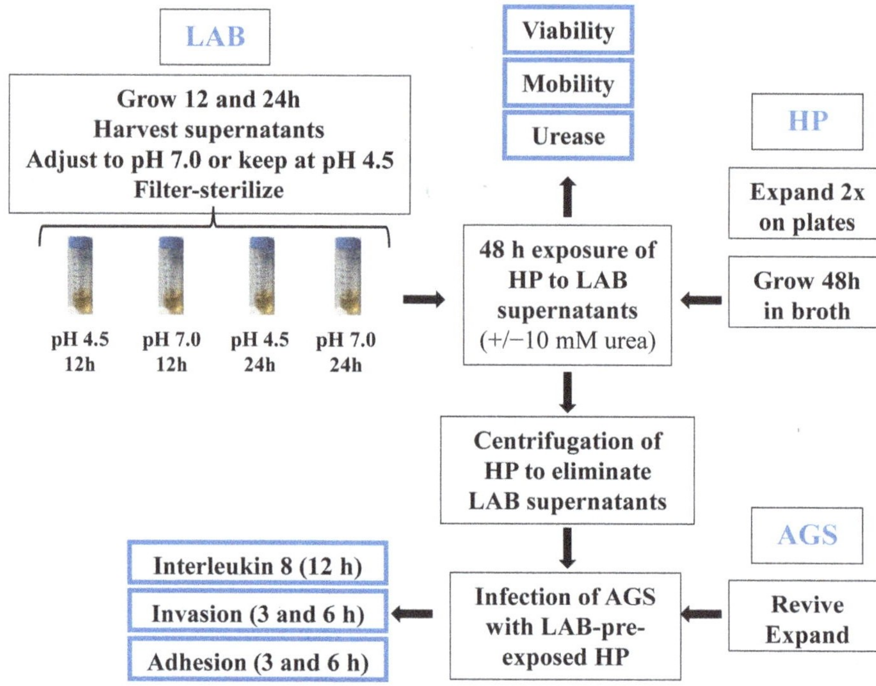

Fig. 1 Schematic representation of experimental layout and all the assays described in this manuscript

(18 × 150 mm, 27 mL capacity), and incubate for 24 h at 37 °C in anaerobic conditions without shaking.

3. After 24 h, transfer 1 mL of each LAB culture to 24 mL of mMRS (i.e., 1/25 dilution) in a 125 mL Erlenmeyer flask and incubate for 12 h, without shaking as above (*see* **Note 3**).
4. After 12 h of incubation, split each 25 mL of the four LAB cultures in half (12.5 mL each). One aliquot of each LAB is incubated for an additional 12 h, and the other is centrifuged (4450× *g* for 20 min at 25 °C). The supernatant is harvested, split in half (6.25 mL), with one aliquot being filter sterilized (using 0.2 μm filters) and stored as native LAB supernatant (pH 3.8–4.5, measured using a pH meter), and the other aliquot being adjusted to pH 7 with 1 M NaOH (3–5 μL depending on LAB native pH), filter sterilized and stored as neutralized LAB supernatant. After the additional 12 h of incubation (total 24 h), each of the 12.5 mL of the four LAB cultures is processed as described above. LAB supernatants are stored at −80 °C for 2–3 weeks.
5. Prepare 6.25 mL of the following controls (*see* **Note 6**). To produce the LA control, acidify modified mMRS broth to pH 4.5 with ~10 μL of DL-lactic acid solution (ACS grade, sold at 85.0–90.0% purity, 9.4–10 M, BioShop). To produce the LN control, acidify 6.25 mL of mMRS broth to pH 4.5

Table 1
Number of CBA plates required to grow one HP strain

Step	Revival	First expansion	Second expansion	Total
# of plates	1	1	5	7

with DL-lactic acid as above, and then neutralize back to pH 7.0 with 1 M NaOH (~15 μL). Store all controls for the same period of time as LAB supernatants.

3.2 HP Growth

The number of CBA plates needed to grow one HP strain necessary to test four LAB is indicated in Table 1 for each step of the process.

1. Prepare seven CBA plates, and allow to dry in BSC or on the bench prior to use.
2. Transfer four loopfuls (of 10 μL each) of frozen HP stock to individual quadrants of a CBA plate. Using a loop, spread the liquid of each spot into a ~ 0.5 cm circle (*see* **Note 7**). To test one HP strain with four LAB strains, prepare one revival HP plate.
3. Incubate 2 days at 37 °C under microaerobic atmosphere (tri-gas incubator unless specified otherwise; *see* **Note 8**). Proceed to the first expansion.
4. Use a cotton swab presoaked in BHI to harvest HP from one of the four spots above and spread the bacteria across a new CBA plate (one HP plate per four LAB to be tested), swabbing in two directions for complete coverage. Incubate 2 days as above.
5. Use a cotton swab presoaked in BHI to transfer HP from the first expansion plate above to a microcentrifuge tube containing 1 mL BHI. Transfer 50 μL of the HP suspension obtained to 950 μL saline in a 1.5 mL spectrophotometer cuvette (1/20 dilution). Read OD_{600} and adjust the rest of the sample to OD_{600} 0.2 with BHI. Make a minimum of 600 μL of this suspension to generate 5 s expansion plates necessary to test four LAB strains. Dip a dry swab in the HP suspension adjusted at OD_{600} 0.2 and spread on a CBA plate. Repeat the dip/swab once on the same plate. Repeat the process on four more CBA plates to generate 5 s expansion plates. Incubate 2 days in microaerobic atmosphere.
6. To harvest HP to broth, use a dry swab to transfer bacteria from each second expansion plate to a 2.0 mL microcentrifuge tube containing 1.5 mL BHI supplemented with 10% FBS, 10 μg/mL vancomycin, 5 μg/mL amphotericin B, 5 μg/mL trimethoprim, and 50 μg/mL sodium pyruvate (one tube/plate). Pool the HP suspension from the five tubes into a 50 mL conical tube. Mix and transfer 10 μL HP suspension

to 990 μL saline in a spectrophotometer cuvette and measure OD_{600}. Use the appropriate volume (calculated based on OD_{600}) of the harvested suspension to prepare 30 mL of bacterial suspension at OD_{600} = 1.5 in BHI in a 125 mL Erlenmeyer flask. Although HP yields may vary, it is expected that ~30 mL of 1.5 OD_{600} HP suspension can be produced at this step from the 5 expansion plates made above, and this is enough to prepare one 96-well exposure plate to test four LAB strains.

7. Incubate the flask loosely capped with foil under microaerobic conditions (jar with Campy GasPak) for 48 h at 37 °C with shaking at 100 RPM. Proceed to the HP exposure to LAB supernatant.

3.3 HP Exposure to LAB Supernatant

The HP-LAB supernatant exposure is carried out in triplicates for each LAB strain and for each of the LA, LN, and BHI media controls, using the plate layout shown in Fig. 2. This layout allows to fit each LAB strain harvested at two different time points and triplicates of each test condition (with and without urea).

1. Following incubation of HP in BHI broth (*see* Subheading 3.2), centrifuge the cultures at 4500× *g* for 20 min, resuspend the pellet in 5 mL fresh supplemented BHI, and adjust to OD_{600} to 1.7. Expect to be able to make a minimum of

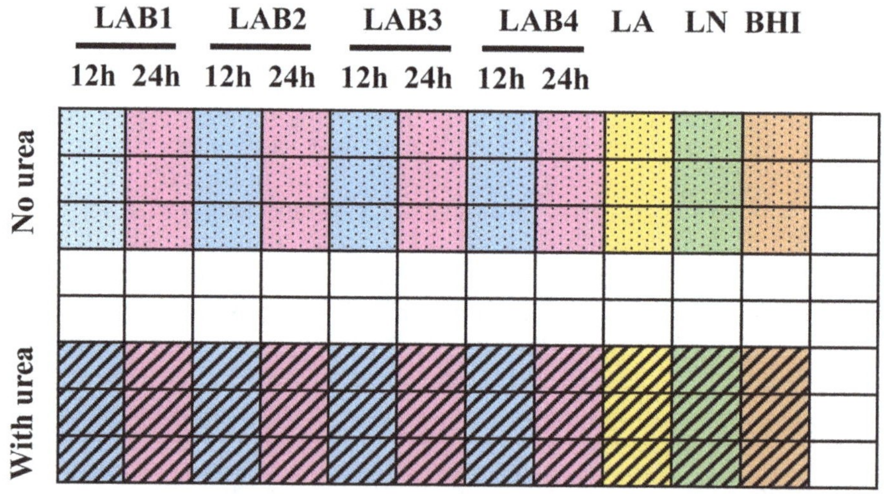

Fig. 2 A schematic representing the sample organization on a 96-well plate to study the effects of four LAB (LAB1 to LAB4) against one HP strain. A full plate allows screening of four LAB strains harvested at 12 h (blue) and 24 h (pink), each with (hatched) and without (dotted) urea, each condition being tested in triplicates. The LA (yellow), LN (green), and BHI (orange) controls are also included on the same plate. White cells are unused. All samples on one plate are tested either at native acidic pH or at neutral pH. After 48h of exposure, the samples from this plate will be used for downstream viability, urease, motility or gastric cell exposure assays

30 mL of HP suspension at OD_{600} 1.7, which is sufficient to test four LAB strains.

2. Split the bacterial suspension in half, with one aliquot receiving urea at a final concentration of 10 mM and the other none.
3. Perform exposure in 96-well plates by combining 150 μL of HP suspension (+/−10 mM urea) with 50 μL of the LAB supernatant or the media control in each well (Fig. 2).
4. Seal the plate in a plastic pouch containing a Campy GasPak (positioned on the side of the plate to avoid overheating samples in the plate) and incubate for 48 h under agitation at 100 RPM (*see* **Note 9**). After 48 h of incubation, read OD_{600}. The OD_{600} reading will be needed for data normalization in subsequent assays. At this stage, phenotypic characterization of LAB supernatant-exposed HP can proceed with the assays described below.

3.4 Processing of LAB Supernatant-Exposed HP for Viability Assay

The viability assay requires six 96-well plates that serve to serial-dilute the LAB-exposed HP, using one 96-well plate per sample row of the exposure plate. Each serial dilution is then spot-plated in triplicates on square CBA plates for enumeration by colony forming unit (CFU) determinations, thus requiring 36 square CBA plates (Table 2, *see* **Note 10**).

1. Prepare 36 square CBA plates (same supplements as above), and allow to dry in BSC or on the bench prior to use.
2. Prepare each dilution plate by introducing 180 μL of 1× PBS in all wells.
3. Using a multichannel pipettor, transfer 20 μL from each well from one row of the exposure plate to the top row (row A) of a dilution plate. Repeat the process until each row of the original exposure plate is set in a dilution plate (6 total).
4. Serially dilute the samples from the top row A tenfold into the subsequent rows by transferring 20 μL of the mix from one row to the next one (dilutions in rows A to H being 10^{-1} to 10^{-8}) of one dilution plate, with mixing by pipetting up and down between each dilution. Tips need to be changed between dilutions for accuracy. Repeat the process on all dilution plates.

Table 2
Number of dilution plates and square CBA plates required to screen the effects of four LAB on one HP strain

# of plates needed	96-well dilution plates	Square CBA spotting plates
Per row of the exposure plate	1	6
Total	6	36

5. Spot-plate all eight serial dilutions in triplicates onto square CBA plates (three columns of eight spots for each sample tested) with samples originating from two exposure wells per square plate (six columns of eight spots/plate).
6. Incubate for 7 days in a microaerobic atmosphere and count the colony forming units (CFU) after 7 days.
7. Process the remaining (180 μL) of the samples of the exposure plates for the other phenotypic assays described in Subheadings 3.5 and 3.6 below.

3.5 Processing of LAB Supernatant-Exposed HP for Motility Assay

This assay is used to determine the ability of LAB-exposed HP to demonstrate flagella-mediated motility in soft agar. It should be done following HP exposure to LAB supernatants (*see* Subheading 3.3). Each sample is stabbed in duplicate into 0.4% soft agar-supplemented BHI plates. As HP motility halos do not exceed 1.5–2 cm, each plate can accommodate six spots; thus only 22 plates are required to process the original 66 exposure samples.

1. Prepare 22 0.4% soft agar supplemented BHI plates (same supplements as above) ~ 6 h before use and keep at room temperature until use.
2. Dip a straight, flame-sterilized, and cooled platinum wire (diameter ~ 1 mm) in the supernatant-exposed HP suspension (~180 μL left-over) to the full depth of the wells, and stab the wire directly into the soft agar plate to the full depth of the media. Repeat the dip/stab process, allowing ~3 cm distance between spots. Flame-sterilize the wire between different LAB samples.
3. Incubate the plates upright in microaerobic atmosphere, and measure the diameter of the bacterial motility halo after 5 days (*see* **Note 11**).

3.6 Processing of LAB Supernatant-Exposed HP for Urease Assay

3.6.1 Principle of the Urease Assay

Urease breaks down urea into CO_2 and ammonia. In the urease assay, the urease present in the samples is first let to react with exogenous urea. The ammonium ions released during the reaction are then quantitatively measured using the Berthelot assay [22, 23], whereby they react with sodium hypochlorite to form a monochloramine, which in turn reacts with two phenols from 2-phenylphenol-nitroprusside to form an indophenol product detectable by its absorbance at 670 nm. This assay is more quantitative and sensitive than the classical phenol red-based assay [24, 25]. Standard curves are prepared using ammonium sulfate, and the urease activity is represented as the fold change in the measured concentration of ammonium ions after exposure to LAB supernatant relative to the appropriate nonexposed control (LN or LA, + vs - urea).

The processing of samples to do the urease assay needs to be done right after the HP-LAB supernatant exposure as above. A

Table 3
Number of urease assay plates required to assess the urease phenotype of four LAB on one HP strain

	LAB-exposed HP sample			
Samples tested	**HP Pellet**	**Supernatant**	**Standards & controls**	**Total**
# test plates needed	2	2	1	5

minimum of two technical replicates is required to assay all standards, controls, and samples, and each test sample is tested for intracellular and extracellular urease activity. Thus, the original exposure plate yields four urease test plates. In addition, a fifth urease test plate is needed to process the standard curve and a control (Table 3).

3.6.2 Procedure of the Urease Assay

1. After withdrawing the necessary amounts for viability and motility testing, the samples from exposure plates are treated as follows. Centrifuge the remaining ~180 μL culture at 4500× *g* and 4 °C for 20 min, transfer the supernatants to a fresh 96-well plate, and keep the original plate now containing only the HP pellets. Wrap both plates tightly with parafilm and store at −20 °C for up to 7 days or − 80 °C for longer periods. Plates with supernatants will be used for extracellular urease assay, and plates with cell pellets will be used for intracellular urease assay.
2. Prepare buffered hypochlorite reagent, sodium phosphate buffer (10 mM, pH 7), 1 M urea in sodium phosphate buffer (pH 7), phenylphenol-nitroprusside reagent (PPN), 5% citrate reagent (pH 7), urease reaction buffer, 1 M urea solution, and a concentrated ammonium sulfate solution (100 mM) as indicated in the materials section above (*see* **Note 5**).
3. Prepare working ammonium sulfate standards solutions daily by diluting the concentrated standard with ddH_2O to obtain concentrations of 0.00, 0.05, 0.50, 1.00, 1.50, 2.00, 2.50, 3.00, and 3.50 mM.
4. Resuspend bacterial pellets in 200 μL of sodium phosphate buffer (10 mM, pH 7). The exposure supernatants are ready to be used without dilution.
5. As one LAB-HP exposure plate yields four urease assay plates, and an additional plate is necessary to process standards and controls (Table 3), prepare five urease test plates. For this, add 90 μL urease reaction buffer to the wells of five 96-well plates.
6. Add 10 μL ammonium sulfate standards to appropriate wells.
7. Add 10 μL bacterial pellet suspensions and supernatants to appropriate plate wells.

8. Incubate, protected from light, for 30 min at RT (*see* **Note 12**).
9. Add 50 μL of phenylphenol-nitroprusside reagent (PPN) followed by 25 μL buffered hypochlorite reagent. These and all the following steps are done in a fume hood, protected from light.
10. Add 100 μL ddH_2O.
11. Incubate 30 min at RT and read OD_{670} with a microplate reader.

3.6.3 Calculation of relative urease activity

1. Samples producing signals greater than the highest standard should be diluted in urease reaction buffer and reanalyzed, then multiplying the concentration found by the appropriate dilution factor.
2. Subtract the mean absorbance value of the standard blank (no ammonium) from all readings. This is the corrected absorbance.
3. Plot the corrected absorbance values of each standard against its concentration, and run a regression fit in the linear portion of the standard curve to obtain the regression curve formula.
4. Back calculate the concentration of ammonium generated by urease activity in the test samples using the regression curve formula and accounting for sample dilutions used (if required).
5. Urease activity expressed in mM of ammonium formed per minute in the total original test samples is calculated as:

$$\text{Urease activity/minute and per exposure well} = \frac{[\text{Ammonium concentration in mM} * \text{Volume of exposure sample in } \mu\text{L}]}{[\text{Reaction time in minutes} * \text{Volume of sample tested in } \mu\text{L}]}$$

Applied to our conditions, this amounts to [2 * Ammonium concentration in mM] / 3].

The OD_{600} of the exposure plate before its 48 h incubation is also plotted along the urease activity data to diagnose whether differences in urease activity correlate with differences in bacterial density at onset of the exposure assay. If needed, the urease data can be normalized to the original bacterial density by dividing the urease data by the OD_{600} values. The ratio of urease activity of a test sample relative to that of its matching, nonexposed control corresponds directly to the ratio of the ammonium concentrations determined for each sample from the standard curves. Finally, the OD_{600} values after 48 h exposure are not used in the urease activity calculation but are plotted alongside the urease data to examine if increased extracellular urease activity and/or decreased intracellular activity correlate with enhanced bacterial lysis (and thus lower OD_{600} after 48 h exposure).

3.7 Tissue Culture of Gastric Cells and their Exposure to HP

This assay investigates the ability of LAB supernatant-exposed HP to induce secretion of pro-inflammatory cytokines (with a focus on IL-8) and to adhere to and invade the gastric cells. To do this, HP needs to be preexposed to the LAB supernatants and pelleted prior to transfer onto the gastric cells (Fig. 2). As a result, a single 96-well plate is needed for each LAB supernatant/HP exposure plate.

3.7.1 AGS-HP Exposure Assay

1. Routinely maintain AGS cells in F-12 K Nutrient Mixture (Kaighn's Modification) supplemented with 10% FBS and penicillin–streptomycin at 37 °C and 5% CO_2. To ensure the highest level of cell viability, expand three times (no more than five) until 20 mL at 2 × 10^5 cells/mL are generated, which is necessary to proceed with assessing a full LAB supernatant/HP exposure plate.
2. Expose HP to LAB supernatant (*see* Subheading 3.3). HP and AGS revival starts at day 1, AGS passages occur at days 3, 5, and 7, HP exposure to LAB supernatants occurs at day 9, and the same day, AGS cells are seeded in a 96-well tissue-culture plate at 2 × 10^5 cells/mL (150 μL/well) and incubated for 48 h. Include wells to accommodate the LN, LA, and BHI media controls of the HP exposure as indicated above in addition to a HP negative control (*see* **Note 13**).
3. After 48 h incubation of LAB-HP exposure, collect 20 μL of each exposure well to assess HP viability as described in Subheading 3.4. Process the remaining exposure samples as follows: Centrifuge the 96-well LAB-HP exposure plate for 20 min at 4500× *g* and 25 °C. Resuspend the resulting bacterial pellets in 200 μL F-12 K media supplemented with 10% FBS only (called, i.e., infection media). Wash the AGS cells (from **step 1**) with 200 μL of 1x PBS, and add 150 μL of the LAB-exposed HP (from 200 μL obtained above) and media control-exposed HP (LA, LN, and BHI) or tissue culture media (infection media) alone to the AGS cell monolayer. Incubate in the CO_2 incubator as required for each assay below.

3.7.2 Adhesion Assay

Incubate the co-culture for 3 and/or 6 h as needed, centrifuge at low speed to pellet the gastric cells without causing cell damage (100× *g* and RT for 5 min), take off the infection media, and remove unattached bacteria by washing the cells four times with 200 μL 1× PBS each time. Lyse the washed AGS-HP-exposed cells with 100 μL saponin (0.08% in 1× PBS), and assess HP viability as described in Subheading 3.4. This will provide the number of HP that have adhered to and entered in the AGS cells. To obtain adhesion values, subtract the invasion data determined below (*see* **Note 14**).

3.7.3 Invasion Assay

Incubate the co-culture for 3 and/or 6 h as needed, centrifuge at 100× *g* and RT for 5 min, and wash the cells four times in PBS as above. Add 200 μL of infection media supplemented with gentamicin at a final concentration of 100 μg/mL per well. Incubate for 30 min in the CO_2 incubator. Remove gentamicin media, and wash the cells four times with PBS as above. Proceed with saponin lysis as above and with spot plating of released bacteria for CFU determinations as described in Subheading 3.4. This will provide the number of HP that have entered (invaded) in the AGS cells (*see* **Note 14**).

3.7.4 IL-8 Measurement

Incubate the co-culture for 12 h at 37 °C as above, centrifuge for 15 min at 1000× *g* and 4 °C. Transfer the supernatants to a new 96-well plate and store sealed at −80 °C until use. Measure the amounts of human IL-8 released by AGS cells in the culture supernatant using an ELISA kit (eBioscience) according to the manufacturer's instructions with the three following modifications: (1) Reconstitute the ELISA standards in F-12 K Nutrient Mixture supplemented with 0.5% BSA. (2) Use 50 μL per well of the 12 h HP-AGS co-culture supernatant (1 well of ELISA per HP-AGS sample well). (3) Standard dilution range should be 0, 250, 500, 750, 1000, 1250, 1500, 1750, and 2000 pmol/mL, and HP-AGS supernatant sample dilution is not preconized since the standard range used best fits the samples.

4 Notes

1. The assays described in this manuscript are labor intensive and require precise timelines and coordination of required components, but they are very robust and are useful for screening multiple LAB secretions for the identification of novel anti-HP agents.
2. Freezer stocks are not generated from mixed freeze-dried bacteria since growth may alter the ratio of the various strains; thus freezer stocks would not be reflective of the original formulation proportions.
3. The switch to mMRS instead of MRS media is necessary to support HP viability in subsequent HP exposure steps.
4. Trimethoprim, vancomycin, and amphotericin B are added in HP media as "background" antibiotics to prevent the growth of other bacteria or fungi during the lengthy incubation necessary to grow HP.
5. For safety, make and use this solution under a fume hood.
6. The LA and LN controls are critical to demonstrate that the activity of LAB secretions is pH-dependent, being mostly active

under acidic pH (for the set of 25 LAB screened so far) and to demonstrate that the acidity itself is not responsible for the observed activity. Both controls comprise LAB growth media acidified with lactic acid to mimic LAB growth. They are then used as such (LA) or after neutralization (LN).

7. HP growth is density-dependent. No HP growth will be observed if the original freezer inoculum is spread thin on the plate or if the expansion from the first revival plate involves excessive dilution of the inoculum. We thus recommend proceeding with a progressive expansion from the freezer stock revival on a very small surface to one full plate and then from that one plate to two to five plates as needed for the actual experiment.
8. The timeline of HP growth is critical. HP forms coccoids upon extensive incubation, even in optimal conditions, and while coccoids are considered viable, they are not culturable. Our timeline is optimized to avoid producing coccoids, and routine examination under the microscope (oil immersion) allows exclusion of cultures that contain more than 10% of coccoids.
9. While HP hardly grows in broth, the 48 h timeline for exposure in broth is optimized to limit coccoid formation and to allow sufficient time for potential transcriptional effects form the LAB secretions on HP to manifest into phenotypic effects (turnover of surface components, synthesis of new virulence factors, metabolic adaptation, etc.).
10. As HP does not grow readily in broth nor lyzes following exposure to LAB supernatants, all viability assays require colony forming unit (CFU) determinations after exposure.
11. Only the motility halo seen within the agar represents motility. Surface-grown inoculum should not be included in the determination of the motility halo diameter. Longer incubation time may increase the bacterial density within the motility halo, allowing easier detection. However, the motility halo diameter rarely expands after 5 days of incubation, probably due to changes in agar rigidity as plates dry over time if a high incubator humidity is not maintained.
12. This step represents the reaction time during which the urease present in the samples can produce ammonium ions. This time may be extended if limited activity is observed.
13. Anticipation of the timeline is critical as synchronization of all active players (exposed HP and live gastric cells in sufficient amounts and seeded in test wells) is key to ensure proper activation of gastric cells by HP.
14. As adhesion and invasion only represent less than 10% and 1% of the inoculum in these experiments, respectively, it is not

necessary to make and spot-plate many tenfold serial dilutions for the LAB-treated controls. Diluting the LAB-exposed samples to 10E4 is usually sufficient.

Acknowledgments

This work was funded via operating grants from the Canadian Institute for Health Research (ARB-183980, 2022) and from Western University (Western Strategic Support-CIHR bridge 2022) to Dr. Creuzenet, as well as via scholarships jointly supported by the Mathematics of Information Technology and Complex Systems program (MITACS) and Lallemand Health Solutions Inc. (Montreal) to post-doctoral fellows S. Whiteside (IT10085), M. Mohiuddin (IT13550), and J. W. Marbou Takougoum (IT36971) and to MSc student D. Carroll (IT32278).

References

1. Wang F, Meng W, Wang B, Qiao L (2014) *Helicobacter pylori*-induced gastric inflammation and gastric cancer. Cancer Lett 345:196
2. Chey WD, Howden CW, Moss SF et al (2024) ACG clinical guideline: treatment of *Helicobacter pylori* infection. Am J Gastroenterol 119: 1730–1753. https://doi.org/10.14309/ajg.0000000000002968
3. Fallone CA, Moss SF, Malfertheiner P (2019) Reconciliation of recent *Helicobacter pylori* treatment guidelines in a time of increasing resistance to antibiotics. Gastroenterology 157:44
4. Homan M, Jones NL, Bontems P et al (2024) Updated joint ESPGHAN/NASPGHAN guidelines for management of *Helicobacter pylori* infection in children and adolescents (2023). J Pediatr Gastroenterol Nutr 79:758. https://doi.org/10.1002/jpn3.12314
5. Aldhaleei WA, Wallace MB, Harris DM, Bi Y (2024) *Helicobacter pylori*: a concise review of the latest treatments against an old foe. Cleve Clin J Med 91:481–487
6. Ho JJC, Navarro M, Sawyer K et al (2022) *Helicobacter pylori* antibiotic resistance in the United States between 2011 and 2021: a systematic review and meta-analysis. Am J Gastroenterol 117:1221
7. Yanhui Y, Jing X, Fangbing L et al (2024) Global primary antibiotic resistance rate of *Helicobacter pylori* in recent 10 years: a systematic review and meta-analysis. Helicobacter 29
8. Tacconelli E, Carrara E, Savoldi A et al (2018) Discovery, research, and development of new antibiotics: the WHO priority list of antibiotic-resistant bacteria and tuberculosis. Lancet Infect Dis 18:318. https://doi.org/10.1016/S1473-3099(17)30753-3
9. Chakravarty K, Gaur S (2019) Role of probiotics in prophylaxis of *Helicobacter pylori* infection. Curr Pharm Biotechnol 20:137. https://doi.org/10.2174/1389201020666190227203107
10. Qureshi N, Li P, Gu Q (2019) Probiotic therapy in *Helicobacter pylori* infection: a potential strategy against a serious pathogen? Appl Microbiol Biotechnol 103:1573
11. Yu HJ, Liu W, Chang Z et al (2015) Probiotic BIFICO cocktail ameliorates *Helicobacter pylori* induced gastritis. World J Gastroenterol 21:6561. https://doi.org/10.3748/wjg.v21.i21.6561
12. Kiattiweerasak A, Aumpan N, Chonprasertsuk S et al (2023) Efficacy and safety of *Lacticaseibacillus rhamnosus R0011* and *Lactobacillus helveticus R0052* as an adjuvant for *helicobacter pylori* eradication: a double-blind, randomized, placebo-controlled study. Front Gastroenterol 2. https://doi.org/10.3389/fgstr.2023.1245993
13. Verdu EF, Bercik P, Xian XH et al (2008) The role of luminal factors in the recovery of gastric function and behavioral changes after chronic *Helicobacter pylori* infection. Am J Physiol Gastrointest Liver Physiol 295:G664. https://doi.org/10.1152/ajpgi.90316.2008
14. Johnson-Henry KC, Mitchell DJ, Avitzur Y et al (2004) Probiotics reduce bacterial

colonization and gastric inflammation in *H. pylori*-infected mice. Dig Dis Sci 49:1095. https://doi.org/10.1023/B:DDAS.0000037794.02040.c2

15. Westerik N, Reid G, Sybesma W, Kort R (2018) The probiotic *Lactobacillus rhamnosus* for alleviation of *Helicobacter pylori*-associated gastric pathology in East Africa. Front Microbiol 9:1873
16. Kabir AMA, Aiba Y, Takagi A et al (1997) Prevention of *Helicobacter pylori* infection by *Lactobacilli* in a gnotobiotic murine model. Gut 41:49. https://doi.org/10.1136/gut.41.1.49
17. Zhang L, Su P, Henriksson A et al (2008) Investigation of the immunomodulatory effects of *Lactobacillus casei* and *Bifidobacterium lactis* on *Helicobacter pylori* infection. Helicobacter 13:183. https://doi.org/10.1111/j.1523-5378.2008.00595.x
18. Whiteside SA, Mohiuddin MM, Shlimon S et al (2021) In vitro framework to assess the anti-*Helicobacter pylori* potential of lactic acid bacteria secretions as alternatives to antibiotics. Int J Mol Sci 22. https://doi.org/10.3390/ijms22115650
19. Liévin Le Moal V, Fayol-Messaoudi D, Servin AL (2013) Compound(s) secreted by *Lactobacillus casei* strain *Shirota YIT9029* irreversibly and reversibly impair the swimming motility of *Helicobacter pylori* and *Salmonella enterica serovar Typhimurium*, respectively. Microbiology (United Kingdom) 159:1956. https://doi.org/10.1099/mic.0.067678-0
20. Song H, Zhou L, Liu D et al (2019) Probiotic effect on *Helicobacter pylori* attachment and inhibition of inflammation in human gastric epithelial cells. Exp Ther Med 18:1551. https://doi.org/10.3892/etm.2019.7742
21. Chen YH, Tsai WH, Wu HY et al (2019) Probiotic *Lactobacillus spp.* act against *Helicobacter pylori*-induced inflammation. J Clin Med 8: 10.3390/jcm8010090
22. Searle PL (1984) The Berthelot or indophenol reaction and its use in the analytical chemistry of nitrogen: a review. Analyst 109:549. https://doi.org/10.1039/AN9840900549
23. Rhine ED, Mulvaney RL, Pratt EJ, Sims GK (1998) Improving the Berthelot reaction for determining ammonium in soil extracts and water. Soil Sci Soc Am J 62:473. https://doi.org/10.2136/sssaj1998.03615995006200020026x
24. Clyne M, Labigne A, Drumm B (1995) *Helicobacter pylori* requires an acidic environment to survive in the presence of urea. Infect Immun 63:1669. https://doi.org/10.1128/iai.63.5.1669-1673.1995
25. Merkx-Jacques A, Obhi RK, Bethune G, Creuzenet C (2004) The *Helicobacter pylori flaA1* and *wbpB* genes control lipopolysaccharide and flagellum synthesis and function. J Bacteriol 186:2253. https://doi.org/10.1128/JB.186.8.2253-2265.2004

Chapter 7

Assessing the Relative Affinities of Bacterial Lectins for Sugars

Robert Eves and Peter L. Davies

Abstract

In this method, a bacterial lectin is allowed to bind to its target polysaccharide in the form of a matrix suspension that can be recovered by centrifugation or filtration. After an initial test to show that the lectin is fully bound to the matrix, a measured amount of competing free sugar is added to the matrix resuspended in buffer. After mixing, the suspension is centrifuged, and the protein content of the supernatant fraction is measured before being returned to the matrix with the addition of another aliquot of sugar. The cycle of mixing, centrifugation, and quantification of protein (lectin) in the supernatant fraction is continued until the released protein reaches plateau levels. The procedure is repeated with a variety of different sugars to compare their efficiencies at binding and displacing the lectin from its polysaccharide ligand. This procedure is illustrated here using a PA14-type bacterial lectin that has affinity for fucosylated glycans and will, as a result, bind to dextran-based resins like Superdex 200 (S200). A second example features the release of *Escherichia coli* maltose-binding protein from amylose resin by mono- and disaccharides. The purpose of this assay is to determine the relative binding affinity of different sugars for a bacterial lectin.

Key words Adhesins, Bacterial lectins, Biofilm, Competition assay, Dextran beads, Glycan, Polysaccharides, Protein binding, Sugars

1 Introduction

Bacteria have developed adhesins to bind various surfaces where there is a nutrient-rich environment [1–3]. There they can form stable colonies called biofilms [4] where the bacteria are encased in a matrix of polysaccharides, proteins, and DNA [5, 6]. Biofilms offer protection from predation [7] and shear forces, and a chance to benefit from the lower energy costs of a sessile lifestyle. Bacteria in biofilms are difficult to remove and are less susceptible to biocides and antibiotics making this a crucial area for study [8, 9].

The initial anchoring of bacteria to surfaces by their adhesins and subsequent attachment to biofilm matrices can occur through lectin domains of many different types [10–13]. Here the lectin's affinity for sugars within glycans or polysaccharides will determine

Inka Brockhausen (ed.), *Dynamics of Bacteria-Mucus Interactions*, Methods in Molecular Biology, vol. 2942, https://doi.org/10.1007/978-1-0716-4627-4_7,

which surfaces the bacteria can colonize [14]. The ability to recognize which sugars are bound by a particular bacterial lectin is important for developing reagents to block bacterial biofilm infections. In this chapter, we illustrate a simple, inexpensive method for determining the sugar specificity of bacterial lectins by using the example of the widespread PA14 bacterial lectin domain that binds to dextrans [15, 16]. The dextran-bound lectin is differentially competed off the matrix by a variety of different sugars. This method is adaptable to any situation where a lectin has affinity for its ligand on a solid support that can be easily recovered by centrifugation or filtration.

To illustrate this method, a His-tagged PA14 domain from *Marinobacter nauticus* (previously named *Marinobacter hydrocarbonoclasticus*) fused to GFP (GFP-*Mn*PA14) is recombinantly expressed in *Escherichia coli* and purified by sequential chromatography on Ni-NTA resin and an anion-exchange column (Q-Sepharose) [15]. After dialysis of the protein into buffer containing $CaCl_2$, an aliquot is bound to equilibrated dextran resin and then washed by resuspension and centrifugation to remove unbound protein. At this point incremental amounts of a sugar are added with mixing, and the A280 nm values of the supernatant recovered by centrifugation are measured to determine how much protein is dissociated from the dextran resin. After each measurement the supernatant is returned to the resin and the cycle is repeated several more times with additional sugar aliquots. When these measurements are plotted, the A280 nm values approach a plateau indicating the effectiveness of that sugar in displacing the lectin from the resin. From these curves apparent dissociation constants are determined.

In another example of this method, maltose-binding protein (MBP) from *E. coli* is bound to amylose beads and then competed off by various di- and monosaccharides. MBP is normally secreted into the periplasm of *E. coli* [17] where it functions to assist in the capture and uptake of the disaccharide, maltose [18]. Maltose has two glucose moieties joined through an α1:4 linkage.

2 Materials

2.1 Special Equipment

1. FPLC apparatus.
2. Centrifuge for 1-L cultures capable of 2500× *g*.
3. Centrifuge capable of reaching 31,000× *g* for a 50 mL sample volume Micro-centrifuge and Eppendorf tubes.
4. UV-spectrophotometer and quartz cuvette (or 96-well plate).
5. Plotting software (e.g., Sigma plot or GraphPad Prism).
6. Sonicator or similar method of bacterial cell lysis.

7. SDS-PAGE apparatus.
8. Vortexer with multi-tube capacity.
9. Nutator or similar mixing apparatus.

2.2 Reagents

1. LB Miller broth and 1.5% agar supplemented with 50 μg/mL kanamycin.
2. Bacterial lectin gene (optional GFP-tag) in a bacterial expression plasmid in *E. coli* BL21 DE3.
3. Maltose-binding protein gene in a bacterial expression plasmid in *E. coli* BL21 DE3.
4. Isopropyl ß-D-1-thiogalactopyranoside (IPTG).
5. Lysis buffer: 50 mM Tris-HCl pH 9.0, 500 mM NaCl, and 2 mM $CaCl_2$, including 1 EDTA-free protease inhibitor tablet for bacterial lysate (Pierce, Thermo Scientific).
6. Ni-buffer: 50 mM Tris-HCl pH 9.0, 500 mM NaCl, 2 mM $CaCl_2$, and 5 mM imidazole.
7. Elution buffer: 50 mM Tris-HCl pH 9.0, 500 mM NaCl, 2 mM $CaCl_2$, and 250 mM imidazole.
8. Column of Ni-NTA resin (25 mL).
9. Nickel sulfate.
10. 10% Tris-Tricine SDS-PAGE gels or equivalent [19].
11. Q16/10 Q-Sepharose anion-exchange column (GE HealthCare) or equivalent.
12. Buffer A: 50 mM Tris-HCl pH 9.0, 2 mM $CaCl_2$.
13. Buffer B: 50 mM Tris-HCl pH 9.0, 1 M NaCl, 2 mM $CaCl_2$.
14. Dialysis tubing 3.5 K cutoff.
15. Superdex 200 resin (GE HealthCare).
16. Amylose resin (Bioworld).
17. Sugars to test in competitive binding assays.

3 Methods

3.1 Cloning the Bacterial Lectin Gene into the Bacterial Expression Vector

1. DNA encoding the *Mn*PA14 sequence is codon optimized for protein production in *E. coli* and synthesized by GeneART (Thermo Fisher) (*see* **Note 1**). Following synthesis, the gene is inserted into the pMA plasmid.
2. Subsequently, the gene is excised from this plasmid using NdeI and XhoI restriction enzymes to cut at the 5′- and 3′-ends of the gene, respectively.
3. The excised gene is then ligated into a pET28a plasmid containing a codon-optimized green-fluorescent protein (GFP)

gene positioned to encode a protein with GFP fused to the N terminus of *Mn*PA14 (*see* **Note 2**).

4. The ligation process uses T4 DNA ligase in a total reaction volume of 20 μL, with a 4:1 ratio of insert to vector and 1 μL of the ligase. This reaction proceeds for 30 min, after which the construct is transformed into chemically competent *E. coli* BL21.

3.2 Expression of the MnPA14 Bacterial Lectin

1. The process begins with streaking *E. coli* BL21 that harbors the GFP-bacterial lectin fusion gene onto an LB agar plate supplemented with 50 μg/mL kanamycin. Colonies become visible after incubation for 18 h at 37 °C (Fig. 1).
2. A single colony is selected and introduced into 100 mL of LB Miller broth containing 50 μg/mL kanamycin within a 250-mL Erlenmeyer flask. This flask is then placed in a 37 °C incubator and shaken at 220 rpm for 18 h.
3. Following this initial culture, two 2-L Erlenmeyer flasks containing 1 L LB Miller broth and 50 μg/mL kanamycin are inoculated with 25 mL of the overnight bacterial culture. The flasks are returned to the incubator and shaken at 220 rpm for approximately 3–4 h or until the OD600 nm reading lies within the range of 0.6–0.9. Subsequently, the flasks are transferred to an open-air shaker at room temperature and shaken

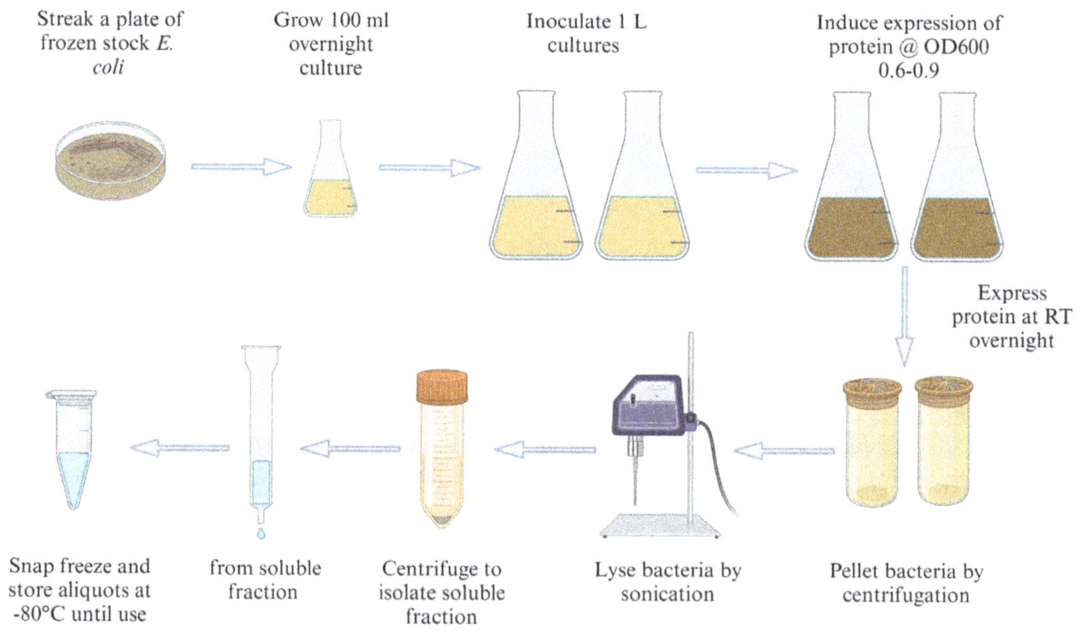

Fig. 1 Bacterial lectin production and purification. The flow chart illustrates the steps required for bacterial lectin purification. (Illustration made using BioRender)

for 1 h. Bacterial culture (1 mL) is taken for SDS-PAGE analysis (pre-induction sample).

4. Induction is initiated with 1 mM IPTG, and the shaking continues at room temperature for an additional 20 h. Again, 1 mL of the bacterial culture is taken for SDS-PAGE analysis (post-induction sample).

3.3 Purification of Protein by Ni-NTA Resin

1. Retrieve the cells through centrifugation using 1-L bottles in a swinging bucket rotor at 2500× *g* for 25 min at 4 °C. Discard the supernatant. Resuspend the resulting pellets in 50 mL of lysis buffer (50 mM Tris-HCl pH 9.0, 500 mM NaCl, and 2 mM $CaCl_2$, containing 1 EDTA-free protease inhibitor tablet (Pierce, Thermo Scientific). Maintain the lysate on ice throughout the procedure.
2. Subject the lysate to sonication at 70% amplitude for 30 s, repeated six times with 5-min cooling intervals between each session (*see* **Note 3**). Following sonication, centrifuge the lysate at 31,000× *g* for 30 min (*see* **Note 4**).
3. Meanwhile, prepare 10 mL of Ni-NTA resin by equilibrating it with Ni-buffer (50 mM Tris-HCl pH 9.0, 500 mM NaCl, 2 mM $CaCl_2$, and 5 mM imidazole) in a 25-mL column. Place the resin in a 500-mL flask along with 200 mL of Ni-buffer and 50 mL of the lysate supernatant obtained from the centrifugation. Gently stir the resin with a magnetic stir bar for 20 min, and then allow it to settle for an additional 20 min.
4. Decant the supernatant to a new flask, preserving it in case a considerable amount of target protein remains unbound, necessitating another round of adsorption to Ni-NTA resin. Reintroduce the resin back into the column and wash it with approximately 60 mL of buffer.
5. Elute the His-tagged protein with 25 mL of Ni-buffer containing 250 mM imidazole. Perform SDS-PAGE on samples obtained from pre-induced bacteria, post-induced samples, lysate pellets, lysate supernatants, flow-through from the resin, and the elution from the resin.

3.4 Purification of Protein by Q-Sepharose Anion Exchange (Optional)

1. Confirm the protein's presence in the Ni affinity column eluate by SDS-PAGE. If there are residual contaminating protein bands in the eluate, dialyze this into Buffer A (50 mM Tris-HCl pH 9.0, 2 mM $CaCl_2$) for anion-exchange chromatography on a Q-Sepharose column in the FPLC system (*see* **Note 5**).
2. Two buffers used for salt gradient elution, Buffer A (50 mM Tris-HCl pH 9.0, 2 mM $CaCl_2$) and Buffer B (50 mM Tris-HCl pH 9.0, 1 M NaCl, 2 mM $CaCl_2$), are prepared and filtered prior to use.

3. The Q-Sepharose column is initially equilibrated with Buffer A at a flow rate of 2 mL/min before introducing the protein sample.
4. Following loading, the column is washed with two column volumes of Buffer A (retained as flow-through).
5. The protein is eluted from the column when Buffer B is gradually introduced at a rate of 1% per min, and the eluate is collected in 3-mL fractions.
6. SDS-PAGE analysis of the fractions is used to confirm the presence of the bacterial lectin protein. The peak fractions containing the protein are subsequently dialyzed into a buffer of 50 mM Tris-HCl (pH 9.0), 150 mM NaCl, and 2 mM $CaCl_2$. The dialyzed fractions are aliquoted and promptly flash-frozen for storage at −80 °C until needed.

3.5 Sugar-Binding Assays

1. To assess the lectin's affinity for dextrans, the protein is loaded onto either an S200 or S75 Superdex column using an FPLC apparatus (AKTA start). The column is initially equilibrated with protein buffer before introducing <5 mL of protein from the previous chromatography. Wash the column with equilibration buffer until the total volume is reached. If the protein has not eluted at this stage, it indicates adherence to the resin (*see* **Note 6**). Elution can be achieved by introducing a gradient of 1 M glucose in the protein buffer at a rate of 1% per minute. Alternatively, if the protein is calcium-dependent, 10 mM EDTA can be used for step elution without requiring a gradient. The protein is then dialyzed into the protein buffer to eliminate glucose or EDTA.
2. Superdex 200 (S200) resin is equilibrated in a buffer of 50 mM Tris-HCl pH 9.0, 150 mM NaCl, and 2 mM $CaCl_2$. This is accomplished by adding 300 μL of a 50% slurry of S200 resin stored in 20% ethanol (150 μL bed volume) to a 1.5 mL microfuge tube, which is then centrifuged at 6600× *g* for 2 min. The supernatant is removed, and the resin is resuspended with 1 mL of buffer, followed by centrifugation and removal of the supernatant. This resin washing process is repeated two more times (*see* **Note 7**) (Fig. 2).
3. A milliliter of 1 mg/mL bacterial lectin is added to the washed and equilibrated S200 resin pellet (~150 μL) in the microfuge tube. After 30 s of gentle vortexing and 2 min of incubation on a nutator (rocking platform), the suspension is centrifuged at 6600× *g* for 2 min, and the supernatant is removed. Subsequently, 1 mL of protein buffer is added to the tube. The mixing and centrifugation process is repeated two more times.

   A UV spectrophotometer is used to measure the $A280_{\text{nm}}$ of the protein in a quartz cuvette. Blank the spectrophotometer using protein buffer. The $A280_{\text{nm}}$ of the supernatant from the

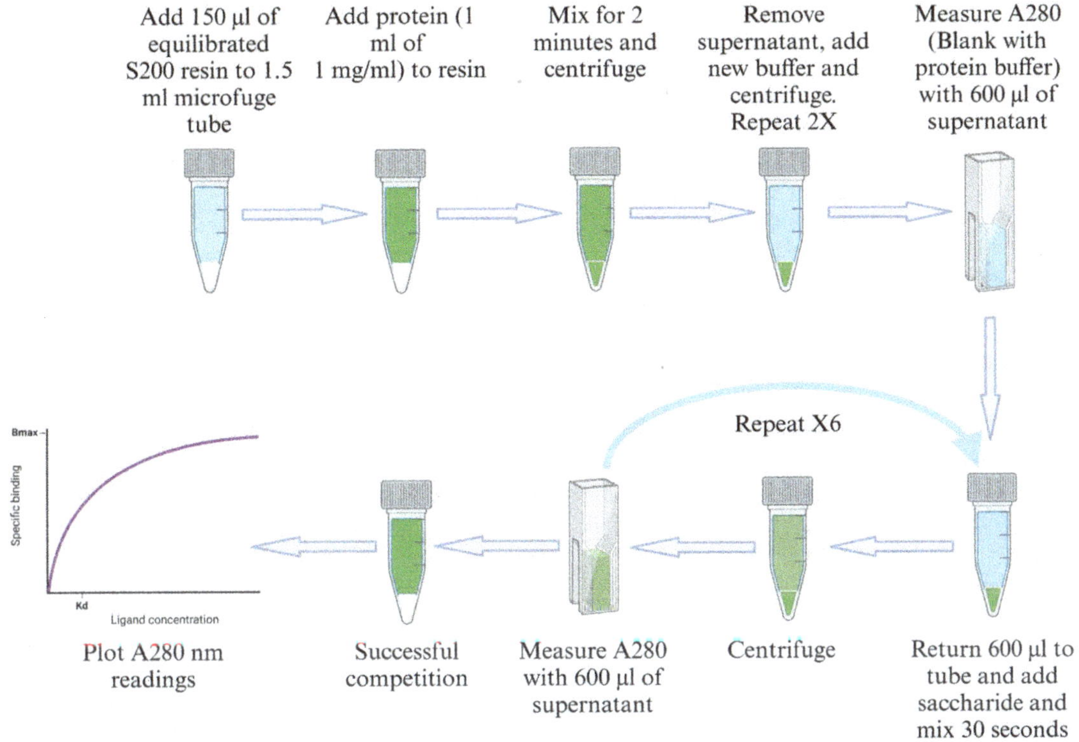

Fig. 2 Sugar binding assay. This flow chart represents the steps involved in performing the sugar competitive binding assay. (Illustration made using BioRender)

last wash of the lectin-bound S200 beads is measured and subtracted from subsequent A280 readings as a background reading for leaching of protein from the resin.

4. At this point, 1.67 µmoles (3 µL of 555 mM) sugar are added to the tube, followed by a 30-s vortexing and centrifugation for 2 min at 6600× *g*. The A280 $_{nm}$ of the supernatant is measured as before and returned to its tube.

5. **Step 4** is repeated six times, followed by a final addition of 5 µmoles (9 µL of 555 mM) of sugar.

   The A280 $_{nm}$ readings for each sugar are plotted using Sigma plot, GraphPad Prism, or other suitable software. The A280 $_{nm}$ of each reading is normalized to the maximum released protein value (Bmax). Data points are collected in triplicate, with error bars representing the standard deviation (Fig. 3). Data are fitted with nonlinear regression using the equation $Y/Bmax = X/(Kd + X)$, where Bmax is the maximum specific binding and Kd is the equilibrium binding constant.

6. To present another example of this method, the displacement of a maltose-binding fusion protein from amylose resin is described. Here the fusion protein is *E. coli* maltose-binding

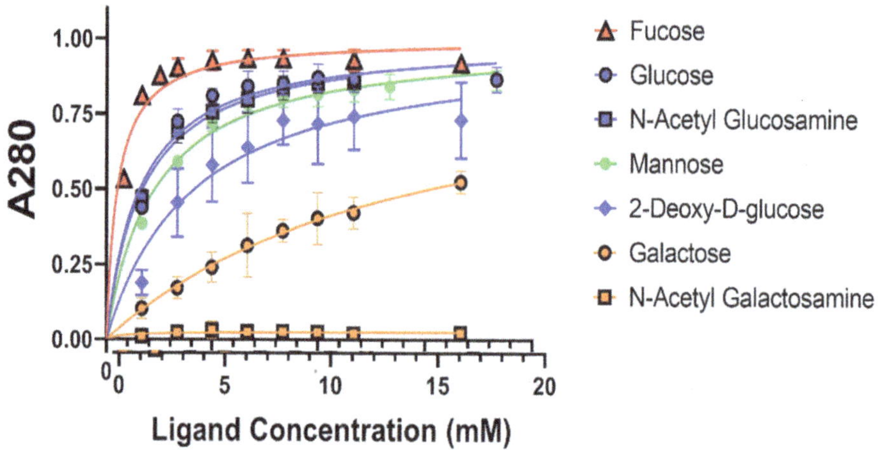

Fig. 3 Dextran-based competitive binding assay to determine relative affinity of *Mn*PA14 for the listed sugars. The progressive removal of *Mn*PA14_GFP from dextran-based resin using increasing concentrations of monosaccharide hexoses is shown by absorbance at 280 nm, normalized to the maximum released protein value (Bmax). Data points were done in triplicate, with standard deviation shown via error bars. Data were fit with nonlinear regression. (This figure is modified from Ref. [15])

protein linked to the C terminus of a 37-kDa human titin fragment, which was a gift from Mathias Bell and Trina Dykstra-McPherson of Queen's University. In the absence of preprepared protein, MBP or a MBP fusion construct can be prepared as a recombinant protein as described for the *Mn*PA14 bacterial lectin (above). **Steps 2–5** described above are repeated with the substitution of the *E. coli* maltose-binding protein for the bacterial lectin and amylose resin for Superdex 200 resin (Fig. 4). Both mono- and disaccharides are tested for their ability to displace the maltose-binding fusion protein from amylose beads.

4 Notes

1. Having your lectin gene codon optimized for expression in *E. coli* and synthesized by a company such as Thermo Fisher (GeneArt) or GenScript can increase your protein yields considerably. Typically, they can also provide the gene ready to express in your vector of choice, saving valuable time and resources.
2. The use of GFP-tagging allows for a reassuring visual display of the protein binding to and elution from, the S200 dextran resin. It is not necessary for the assay providing the lectin you are using contains enough aromatic amino acids for accurate $A280_{nm}$ readings. Alternatively, small quantities of the elutes

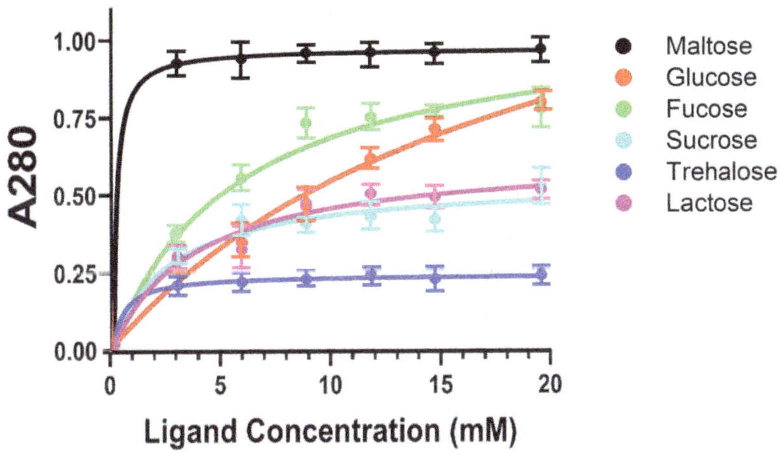

Fig. 4 Amylose-based competitive binding assay to determine relative affinity of MBP-fusion protein for the listed sugars. The progressive removal of *E. coli* maltose-binding protein linked to the titin fragment from amylose resin using increasing concentrations of mono- and disaccharides is shown by absorbance at 280 nm, normalized to the maximum released protein value (Bmax). Data points were done in triplicate, with standard deviation shown via error bars. Data were fit with nonlinear regression

can be assayed for protein by the Bradford or other methods. The GFP tag also offers the opportunity to measure elution by fluorescence.

3. While sonicating the *E. coli* culture, maintain the probe at approximately 1–3 mm from the flask's bottom. Between sonic pulses, remove the lysate and substitute it with a beaker of ice water to cool the probe to minimize heating the lysate. To monitor bacterial lysis progress, prepare 1.5 mL microfuge tubes each containing 1 mL of Bradford reagent. Following each sonic pulse, transfer 2 μL of lysate to the Bradford reagent. Complete bacterial lysis is indicated by no further increase in the Bradford assay color.
4. To ensure equal loading of pellet to supernatant on SDS-PAGE, 50 μL of lysate in a 1.5 mL microfuge tube is centrifuged at 17,000× *g* for 15 min. The supernatant is then transferred to a new tube and the pellet is resuspended in 50 μL of lysis buffer and 3 μL of each is added to the 10-well SDS-PAGE.
5. This step can be excluded if your protein is of high purity after the Ni-NTA purification.
6. This step can be used as an affinity purification step in lieu of Q-Sepharose purification. This will also separate properly folded lectins from any that are poorly folded (which will not bind the column and will elute before the total volume).

7. The equilibration and binding of the protein can be done as a larger batch and aliquoted to 150 μL bed volumes if multiple proteins or sugars are going to be assayed.

Acknowledgments

This work was funded by NSERC Discovery Grant RGPIN-2016-04810 to PLD who holds the Canada Research Chair in Protein Engineering. We are grateful to Mathias Bell and Trina Dykstra-McPherson of Queen's University for the gift of the purified MBP-titin fusion protein.

References

1. Zobell CE (1943) The effect of solid surfaces upon bacterial activity. J Bacteriol 46(1): 39–56. https://doi.org/10.1128/jb.46.1.39-56.1943
2. Tuson HH, Weibel DB (2013) Bacteria-surface interactions. Soft Matter 9(18): 4368–4380. https://doi.org/10.1039/C3SM27705D
3. Berne C, Ducret A, Hardy GG, Brun YV (2015) Adhesins involved in attachment to abiotic surfaces by gram-negative bacteria. Microbiol Spectrum 3(4). https://doi.org/10.1128/microbiolspec.MB-0018-2015
4. Costerton JW et al (1987) Bacterial biofilms in nature and disease. Ann Rev Microbiol 41: 435–464. https://doi.org/10.1146/annurev.mi.41.100187.002251
5. Hoiby N (2017) A short history of microbial biofilms and biofilm infections. APMIS 125(4):272–275. https://doi.org/10.1111/apm.12686
6. Flemming HC et al (2023) The biofilm matrix: multitasking in a shared space. Nat Rev Microbiol 21(2):70–86. https://doi.org/10.1038/s41579-022-00791-0
7. Goh YF et al (2023) Associational resistance to predation by protists in a mixed species biofilm. Appl Environ Microbiol 89(2):e0174122. https://doi.org/10.1128/aem.01741-22
8. Pai L, Patil S, Liu S, Wen F (2023) A growing battlefield in the war against biofilm-induced antimicrobial resistance: insights from reviews on antibiotic resistance. Front Cell Infect Microbiol 13:1327069. https://doi.org/10.3389/fcimb.2023.1327069
9. Schembri MA, Givskov M, Klemm P (2002) An attractive surface: gram-negative bacterial biofilms. Sci STKE 2002(132):re6. https://doi.org/10.1126/stke.2002.132.re6
10. Rostand KS, Esko JD (1997) Microbial adherence to and invasion through proteoglycans. Infect Immun 65(1):1–8. https://doi.org/10.1128/iai.65.1.1-8.1997
11. Rather MA, Gupta K, Mandal M (2021) Microbial biofilm: formation, architecture, antibiotic resistance, and control strategies. Braz J Microbiol 52(4):1701–1718. https://doi.org/10.1007/s42770-021-00624-x
12. Bouckaert J et al (2005) Receptor binding studies disclose a novel class of high-affinity inhibitors of the Escherichia coli FimH adhesin. Mol Microbiol 55(2):441–455. https://doi.org/10.1111/j.1365-2958.2004.04415.x
13. Scharenberg M, Schwardt O, Rabbani S, Ernst B (2012) Target selectivity of FimH antagonists. J Med Chem 55(22):9810–9816. https://doi.org/10.1021/jm3010338
14. Imberty A, Varrot A (2008) Microbial recognition of human cell surface glycoconjugates. Curr Opin Struct Biol 18(5):567–576. https://doi.org/10.1016/j.sbi.2008.08.001
15. Vance TDR et al (2019) Structure and functional analysis of a bacterial adhesin sugar-binding domain. PLoS One 14(7):e0220045. https://doi.org/10.1371/journal.pone.0220045
16. Guo S et al (2021) Structural basis of ligand selectivity by a bacterial adhesin lectin involved in multispecies biofilm formation. mBio 12(2): e00130-21. https://doi.org/10.1128/mBio.00130-21
17. Shuman HA (1982) Active transport of maltose in Escherichia coli K12. Role of the periplasmic maltose-binding protein and evidence for a substrate recognition site in the cytoplasmic membrane. J Biol Chem 257(10): 5455–5461

18. Ames GF (1986) Bacterial periplasmic transport systems: structure, mechanism, and evolution. Annu Rev Biochem 55:397–425. https://doi.org/10.1146/annurev.bi.55.070186.002145

19. Schagger H, von Jagow G (1987) Tricine-sodium dodecyl sulfate-polyacrylamide gel electrophoresis for the separation of proteins in the range from 1 to 100 kDa. Anal Biochem 166(2):368–379. https://doi.org/10.1016/0003-2697(87)90587-2

Chapter 8

Investigating Bacterial Invasion of Highly Differentiated Human Bronchial Epithelial Cell Barriers

Hannah J. Parks and Miguel A. Valvano

Abstract

Confocal microscopy provides information on host–pathogen interactions. To examine infection at the respiratory epithelial barrier, primary bronchial cells differentiated at an air-liquid interface (ALI) can be imaged and reconstructed as a 3D model. In this chapter, we describe a method for immunostaining and fluorescence imaging of well-differentiated bronchial epithelial tissue that has been exposed to the cystic fibrosis respiratory pathogen *Achromobacter xylosoxidans*.

Key words Cystic fibrosis, Air–liquid interface culture, Tight junctions, Transepithelial electrical resistance, ZO-1, Confocal microscopy

1 Introduction

Achromobacter species are opportunistic Gram-negative bacteria that cause acute and chronic, life-threatening infections in immunocompromised individuals. Respiratory infections by *Achromobacter* species occur in people with cystic fibrosis and are difficult to treat due to the bacteria's multidrug resistance [1, 2]. Moreover, chronic infection with *Achromobacter* species in people with cystic fibrosis is associated with a decline in lung function and higher frequency of pulmonary exacerbation [3, 4], including recolonization after lung transplantation [5].

Despite the evidence described in clinical studies, not much is known about how *Achromobacter* interacts with the epithelial tissues lining the bronchia. In this chapter, we provide a method to assess the integrity of well-differentiated human bronchial epithelial tissue during *Achromobacter* infection by measuring transepithelial electrical resistance (TEER) [6]. We also describe a detailed protocol to acquire confocal microscopy images upon immunofluorescent staining of the tight junctions to interrogate the health of the epithelial barrier in the context of infection with *Achromobacter*

Inka Brockhausen (ed.), *Dynamics of Bacteria-Mucus Interactions*, Methods in Molecular Biology, vol. 2942,
https://doi.org/10.1007/978-1-0716-4627-4_8,

[7]. We demonstrate that 3D model reconstruction of the epithelial layer provides clues to evaluate how chronic *Achromobacter* respiratory infections are established. The protocols described can be extended to investigate interactions between epithelial tissue and other respiratory pathogens.

2 Materials

2.1 Bacterial Growth Medium and Strains

1. Mineral-Tryptone-Glycerol (MTG) medium [8]: 0.2% (NH4) 2SO4 (Sigma), 0.6% Na2HPO4 (Sigma), 0.3% KH2PO4 (Sigma), 0.3% NaCl (Sigma), 0.1% KNO3 (Sigma), 1% glycerol (Honeywell), and 0.5% tryptone (Melford Biolaboratories). Tryptone + agar should be prepared and autoclaved separately and MTG salts solution added to warm agar prior to plating (*see* **Note 1**).
2. Stock of appropriate antibiotics at final concentrations of 30–100 μg/ml to select for the recombinant plasmid encoding the mCherry red fluorescent protein (*see* **Note 2**).
3. The bacterium used to demonstrate this method was part of a collection of clinical isolates of *Achromobacter* species from people with cystic fibrosis, which were obtained from the Antimicrobial Resistance and Healthcare Associated Infections Reference Unit of the UK Health Security Agency [8].

2.2 Tissue Culture Medium and Cells

1. Airway epithelial medium supplemented with/without antibiotics (penicillin/streptomycin). For this protocol we have used MucilAir™ medium manufactured by Epithelix (*see* **Note 3**).
2. Well-differentiated primary bronchial epithelial cells (*see* **Note 4**).

2.3 Infection

1. Tabletop centrifuge.
2. Centrifuge (for tissue culture plates).
3. Spectrophotometer.
4. Cell culture incubator.

2.4 Transepithelial Electrical Resistance (TEER)

1. Epithelial Volt/Ohm meter (EVOM®; World Precision Instruments).

2.5 Colony Forming Units

1. Agar, granulated, high strength (Melford).
2. Petri dishes (Sarstedt).

2.6 Immunofluorescent Staining

1. Phosphate-buffered saline (PBS).
2. Triton X-100 (Sigma-Aldrich).
3. Albumin, Bovine Fraction V (BSA) (Melford).
4. Hoechst 33342 staining dye solution (Abcam) (4 μM).
5. ZO-1 monoclonal (Alexa 488) conjugated primary antibody (Thermo Fisher Scientific) (1:100).
6. Anti-mucin 5 AC primary antibody (Abcam) (1:250).
7. Goat anti-Mouse IgG (Alexa 488) (Thermo Fisher Scientific) (1:2000).
8. Goat anti-Mouse IgG (Alexa 647) (Thermo Fisher Scientific) (1:2000).
9. ActinGreen 488 ReadyProbes (Thermo Fisher Scientific) (2 drops/ml).
10. 4% paraformaldehyde (PFA).
11. 14 mM stock of ammonium chloride.
12. Fluoroshield with 1,4-Diazabicyclo[2.2.2]octane (Sigma-Aldrich).

2.7 Imaging Analysis

1. Stellaris SP5 confocal microscope (Leica); LasX software (Leica).

3 Methods

3.1 Infection

1. Streak *Achromobacter* isolate on MTG agar supplemented with 30 μg/mL chloramphenicol and incubate overnight at 37 °C.
2. Prepare a bacterial liquid culture by inoculating a single colony in 5 mL MTG containing 30 μg/mL chloramphenicol.
3. Next morning replace basolateral medium of epithelial tissue with antibiotic-free epithelial cell culture medium, if cells were cultured with antibiotics.
4. Collect 1 mL of overnight bacterial culture into a sterile Eppendorf tube, and wash three times with sterile antibiotic-free epithelial cell culture medium by pelleting and resuspending the bacteria using a tabletop centrifuge.
5. Measure the optical density of the washed bacterial culture using the spectrophotometer at 600 nm.
6. Calculate volume of bacteria needed for infection (*see* **Note 5**):

$$\frac{\text{MOI} \times \text{volume in well} \times \frac{\text{cells}}{\text{mL}}}{\frac{\text{CFU}}{\text{mL}} \times \text{OD600}}.$$

7. Add bacterial suspension to apical surface of the tissues.

8. Add sterile cell culture medium to the uninfected control to act as a blank.
9. Incubate for the desired times at 37 °C, 5% CO_2 incubator.

3.2 TEER Measurements

1. Sterilize EVOM electrode in 70% ethanol for 10 min.
2. Blank in sterile medium.
3. Place electrode in well, ensuring not to disturb the cells, and take a reading. Repeat at by placing the electrode at two additional sites of the well. Calculate the average of the three readings (*see* **Note 6**).
4. Calculate Ohm per centimeter square by multiplying the reading by the area of membrane (see Fig. 1).

3.3 Colony Forming Units (CFUs)

1. At appropriate times postinfection, collect apical and basolateral medium.
2. Conduct serial dilutions of the samples collected in epithelial cell culture medium or PBS as preferred.
3. Spot 10 μL of each sample in triplicate on freshly prepared MTG agar plates supplemented with 30 μg/mL chloramphenicol.
4. Incubate for 2 days at 37 °C.
5. Count colonies from each spot, and calculate the average CFU/mL.

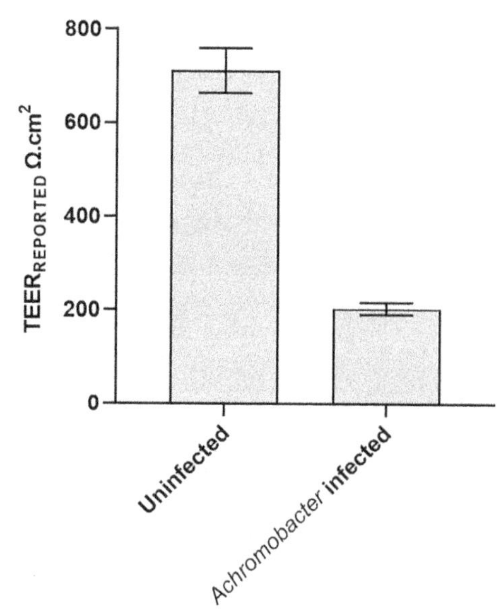

Fig. 1 Example TEER measurement of uninfected tissues vs those infected with an *Achromobacter* clinical isolate

3.4 Immuno-fluorescent Staining

1. Wash the apical surface of the epithelial cells with sterile PBS two times.
2. Fix for 40 min with 4% PFA.
3. Wash apically with sterile PBS three times.
4. Add ammonium chloride to a final concentration of 14 mM and incubate for a minimum of 20 min to stain immediately (*see* **Note 7**).
5. Wash three times with PBS.
6. Permeabilize with 0.1% Triton X-100 for 15 min at room temperature.
7. Wash three times with PBS.
8. Block with 1% BSA for 30 min at room temperature.
9. Add primary antibody, incubate overnight at 4 °C with gentle agitation (*see* **Note 8**).
10. Wash three times with PBS.
11. Add secondary antibody for 2 h at room temperature with gentle agitation.
12. Wash three times with PBS.
13. Incubate with Hoechst 33342 for 40 min in PBS (*see* **Note 9**).
14. Wash three times with PBS.
15. Cut out porous membrane using scalpel.
16. Place membrane down, cells up on microscope slide.
17. Add drop of fluoroshield on cell side, mount coverslip.

3.5 Imaging and Processing

1. Set lasers for each channel to be visualized on the confocal microscope, including bright field. Specific settings will depend on the spectral characteristics of the fluorophores.
2. Start with laser intensity and smart gain both low and gradually increase until signal is clear. (*see* **Note 10**).
3. Set bit depth as high as feasible to achieve maximum number of gray levels. Smaller bit depths can be used to minimize data usage.
4. Adjust the pinhole as required to change the thickness of the optical section.
5. A higher magnification of objective lens is recommended dependent on the size of the sample. We used × 100 for imaging the bacteria.
6. Select points at random across the sample to acquire multiple images.

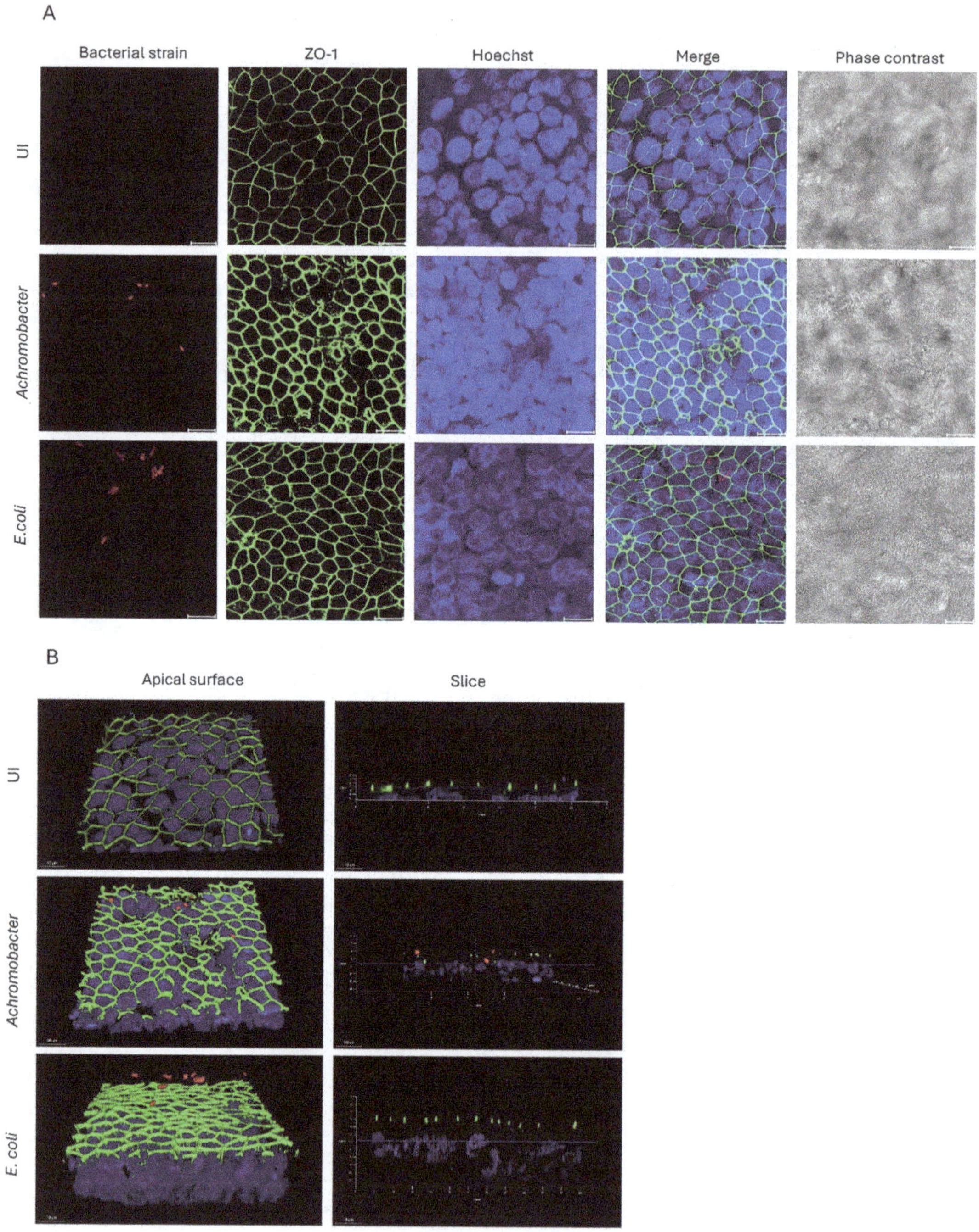

Fig. 2 (**a**) Differentiated bronchial epithelial tissue stained with ZO-1 and Hoechst 33342. Infected with an *Achromobacter* clinical isolate, and an *Escherichia coli* control. Images taken with × 100 magnification on Leica Stellaris SP5 confocal microscope. Scale bar, 10 μm. (**b**) 3D models constructed using the Leica LasX software, with sections shown. *UI* uninfected

7. At each point, acquire a Z stack, a series of slices taken in the Z axis, using a 0.3 μm step size, starting from above the cilia and ending at pores of the plastic membrane.
8. A maximum projection or 3D model of a Z stack can be produced using LASX, FIJI/ImageJ, or other suitable software (*see* **Note 11**) (see Figs. 2, 3 and 4). Interactions between the bacteria and epithelial cells can be studied using image analysis software such as IMARIS (Oxford Instruments).

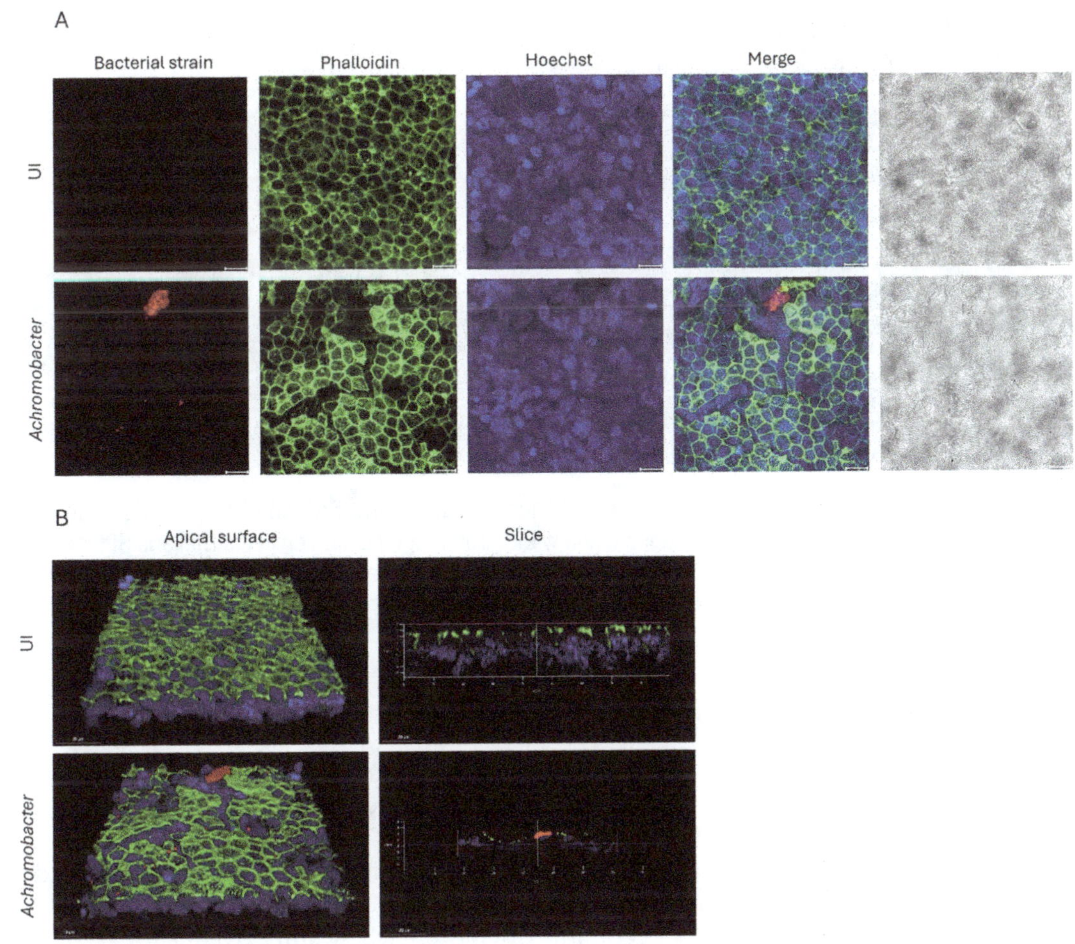

Fig. 3 (**a**) Differentiated bronchial epithelial tissue stained with ActinGreen 488 and Hoechst 33342. Infected with an *Achromobacter* clinical isolate. Images taken with × 100 magnification on Leica Stellaris SP5 confocal microscope. Scale bar, 10 μm. (**b**) 3D models constructed using the Leica LasX software, with sections shown. *UI* uninfected

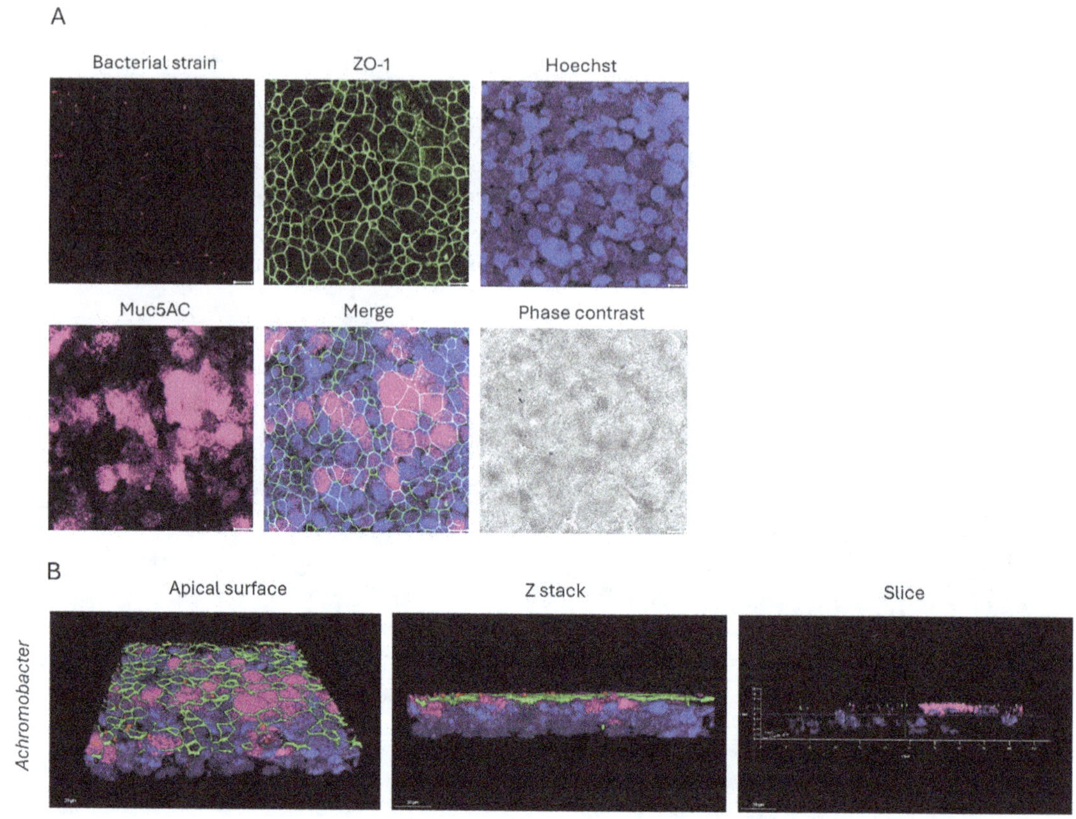

Fig. 4 (**a**) Differentiated bronchial epithelial tissue stained with ZO-1, muc5AC and Hoechst 33342. Infected with an *Achromobacter* clinical isolate. Images taken with × 100 magnification on Leica Stellaris SP5 confocal microscope. Scale bar, 10 μm. (**b**) 3D models constructed using the Leica LasX software, with sections shown. *UI* uninfected

4 Notes

1. The growth medium depends on the bacterial species/strains used for these experiments. Tryptone agar is prepared and autoclaved separately to the remaining media, which is added prior to inoculation with *Achromobacter*. This is due to the media becoming discolored when autoclaved with tryptone. A 10X MG stock can be prepared. The MTG medium has been optimized in our laboratory to grow *Achromobacter* species [8].
2. We routinely used mCherry encoded by the plasmid pJT05 to fluorescently label the bacterial strains [9]. Maintenance of this plasmid requires chloramphenicol, but this can be modified accordingly if another plasmid is used or if the fluorescent

protein-encoding gene is integrated directly into the bacterial chromosome, in which case no antibiotic selection will be required.

3. Popular epithelial cell culture medium includes those manufactured by PromoCell and Stemcell technologies.
4. Well-differentiated primary epithelial cells used for this protocol may be purchased directly from companies (e.g., Epithelix), where they already come prepared in transwell dishes. Alternatively, the cultures can be obtained from clinical sources (e.g., nasal brushes) and differentiated as air–liquid interface cultures in transwell dishes using published protocols [10].
5. Volume of bacteria calculated depends on the cell density.
6. To measure the TEER, EVOM electrodes should be fully immersed in the medium both at the apical and basolateral chambers of the tissue insert.
7. Tissues can be stored in 14 mM ammonium chloride at 4 °C for up to 6 months, with care taken to ensure the cells do not dry out.
8. If staining with two primary antibodies, it is recommended to do this separately over the course of two overnight incubation periods.
9. At this point, cells can be incubated with ActinGreen 488 at the same time as Hoechst 33342, for the same incubation time.
10. Caution is advised with increasing both laser intensity and smart gain if the signal of the fluorophores is low. Increasing laser intensity can cause photobleaching of sample, while increasing the smart gain can amplify background signal.
11. 3D model construction can also be accomplished using IMARIS, which gives the opportunity for detailed image analyses.

Acknowledgments

We thank Dr. Ryan Delaney for technical assistance in confocal imaging and 3D reconstruction and Dr. Joanna Sá-Pessoa for advice with culturing epithelial cells. This work was supported by Cystic Fibrosis Foundation grant VALVAN19G0 to M.A.V.

References

1. Swenson CE, Sadikot RT (2015) *Achromobacter* respiratory infections. Ann Am Thorac Soc 12(2):252–258. https://doi.org/10.1513/AnnalsATS.201406-288FR
2. Gabrielaite M et al (2021) Transmission and antibiotic resistance of *Achromobacter* in cystic fibrosis. J Clin Microbiol 59(4). https://doi.org/10.1128/JCM.02911-20

3. Tetart M et al (2019) Impact of *Achromobacter xylosoxidans* isolation on the respiratory function of adult patients with cystic fibrosis. ERJ Open Res 5(4):00051. https://doi.org/10.1183/23120541.00051-2019
4. Kerem E et al (2023) Clinical outcomes associated with *Achromobacter* species infection in people with cystic fibrosis. J Cyst Fibros 22(2): 334–343. https://doi.org/10.1016/j.jcf.2022.11.001
5. Holm AE et al (2021) Bacterial re-colonization occurs early after lung transplantation in cystic fibrosis patients. J Clin Med 10(6). https://doi.org/10.3390/jcm10061275
6. Srinivasan B et al (2015) TEER measurement techniques for in vitro barrier model systems. J Lab Autom 20(2):107–126. https://doi.org/10.1177/2211068214561025
7. Kuo WT, Odenwald MA, Turner JR, Zuo L (2022) Tight junction proteins occludin and ZO-1 as regulators of epithelial proliferation and survival. Ann N Y Acad Sci 1514(1): 21–33. https://doi.org/10.1111/nyas.14798
8. MacDonald L et al (2023) Polymyxin resistance and heteroresistance are common in clinical isolates of *Achromobacter* species and correlate with modifications of the lipid A moiety of lipopolysaccharide. Microbiol Spectrum 11(1):e0372922. https://doi.org/10.1128/spectrum.03729-22
9. Turton K et al (2023) The *Achromobacter* type 3 secretion system drives pyroptosis and immunopathology via independent activation of NLRC4 and NLRP3 inflammasomes. Cell Rep 42(8):113012. https://doi.org/10.1016/j.celrep.2023.113012
10. Broadbent L et al (2020) Comparative primary paediatric nasal epithelial cell culture differentiation and RSV-induced cytopathogenesis following culture in two commercial media. PLoS One 15(3):e0228229. https://doi.org/10.1371/journal.pone.0228229

Chapter 9

Methods of a Reproducible Mouse Model of *Clostridioides difficile* Infection to Investigate Novel Bacterial-Based Therapies

Katya Douchant, Jill Greenlaw, and Prameet M. Sheth

Abstract

Mice are widely used in science and have played an instrumental role in understanding immune mechanisms, testing cancer therapies, and serving as model systems to better understand physiological processess. Mouse models have also been widely used in infectious disease research including in *Clostridioides difficile* infection (CDI). Mice display similar disease characteristics to human CDI, including susceptibility to disease, clinical symptoms, and pathology, and thus are an ideal in vivo model to investigate treatments. Here we describe a reliable, reproducible model of CDI that allows researchers to investigate the use of prophylactic oral treatments for CDI in mice. Using this model, we report on treatment efficacy using several parameters as a measure of CDI, including survival, bodyweight, colon length, and colonic histology.

Key words Mouse model, *Clostridioides difficile*, Infection, Toxin, Histology

1 Introduction

Animal models have played, and continue to play, an instrumental role in infectious disease research. Animal models replicate the biological processes of disease found in humans in a controlled environment that allows for careful study of host-immune responses, disease pathogenesis, and disease mechanisms and are amenable to genetic manipulation that can be used to interrupt genes to further study host–pathogen interactions [1]. Animal models have contributed immensely by enabling the development of diagnostic tests, investigating interventions (such as vaccines and therapeutics), enhancing our understanding of disease transmission and pathogenesis, ultimately saving countless lives [1]. Here we outline the methods and details of animal models that have been used for the study of *Clostridioides difficile* and outline some of the advantages and disadvantages of each model. As always, the choice

Inka Brockhausen (ed.), *Dynamics of Bacteria-Mucus Interactions*, Methods in Molecular Biology, vol. 2942,
https://doi.org/10.1007/978-1-0716-4627-4_9,

of model should be governed by the research question and study aims.

C. difficile (CD) is a Gram-positive, spore-forming, toxin-producing anaerobic bacterium that is both a human and animal pathogen. CD can colonize and cause disease spontaneously in a variety of species, including horses, pigs, calves, dogs, cats, hamsters, guinea pigs, rats, mice, and rabbits [2]. CD colonization is achieved through disruption of the normal gastrointestinal (GI) microbiota (GIM), such as following antibiotic exposure, creating a permissive environment for its germination and growth [3]. Following colonization, CD releases toxins (TcdA and TcdB) that inactivate Rho family GTPases, causing disorganization of the actin cytoskeleton in GI epithelium followed by cell rounding and cell death [4]. CD infection (CDI) causes symptoms that include diarrhea, colitis, abdominal cramping and pain, and, in severe cases, pseudomembranous colitis, toxic megacolon, and death [5] (Table 1).

The Center for Disease Control has designated CD as an urgent threat to public health, due to its significant impact on patient morbidity, mortality, and healthcare costs (https://www.cdc.gov/antimicrobial-resistance/data-research/threats/index.html). CDI is the leading cause of antibiotic-associated diarrhea, contributing to >500,000 infections and ~30,000 deaths each year in the United States alone [14]. The Public Health Agency of Canada reports that the incidence of CDI is 4.9 cases per 10,000 patient-days, with rates being highest among the elderly and those in hospital [15]. The annual economic burden of CDI in Canada has been estimated at approximately CAD $281 million, largely due to increased duration of hospital stays and costs associated with CDI patient isolation (17.3 days compared to 6.1 days for patients without CDI) [16]. CDI is also associated with a high rate of recurrence, which can contribute to further morbidity, mortality, and prolonged hospital admissions [17]. Following completion of standard-of-care antibiotic therapy, up to 25% of patients develop recurrent CDI (rCDI), which is more difficult to treat and is associated with increased hospitalizations, more severe outcomes, and higher treatment costs than the initial infection [18, 19]. Furthermore, after a first recurrence, there is a higher chance of additional recurrences, where up to 65% of people who have had one instance of rCDI are at risk for subsequent rCDI [20–22]. Given its burden of disease and high rates of treatment failure, there is an urgent need to develop safe, novel therapeutics to control infection.

To investigate novel treatments to control CDI, several mouse models of infection have been developed [6, 7]. Mice are the ideal model to investigate treatment of disease since they share several similarities with human infections such as disease susceptibility, primary niche of colonization of CD, and disease pathogenesis. In these established models, following antibiotic administration, the

Table 1
***In vivo* models for infection**

Animal model	Description	Strengths	Limitations	Examples
Mouse model	Widely used; involves antibiotic-mediated GIM disruption prior to administering (CD) spores to induce CDI	Well-characterized genetics, cost-effective, and widely available	Often succumb to disease within 24–72 h due to a more aggressive disease course, which is rare in humans	C57BL/6 mice receive antibiotic cocktail in drinking water (kanamycin, gentamicin, vancomycin, colistin, metronidazole) for 3 days followed by intraperitoneal injection of clindamycin prior to infection [6] BALB/c mice are orally gavaged with cyclosporin daily for 7 days prior to infection [7]
Hamster model	The first animal model of CDI, involves the administration of CD spores	Highly reproducible	Not fully representative of human disease. CDI in hamsters manifests primarily in cecum compared to humans where CDI primarily affects the colon. Additionally, hamsters do not typically develop diarrhea which is a key hallmark of CDI in humans. (PMID 22555466). CDI in hamsters is lethal which is not representative of human CDI (*see* **Note 1**)	Syrian hamsters are pretreated with vancomycin in drinking water for 7 days prior to infection [8] Golden Syrian hamsters are orally gavaged with clindamycin prior to infection [9]
Rat model	Used less frequently but involves similar infection methods	Allows study of a larger animal physiology;	Less standardized; not as widely used	Wistar rats receive two intramuscular injections per day

(continued)

Table 1 (continued)

Animal model	Description	Strengths	Limitations	Examples
	as mice and hamsters	disease is not fatal which is more similar to human disease		of metronidazole for 3 days prior to infection [10]
Prairie dog model	Involves administration of cefoxitin to induce colitis	Disease course is more chronic than other animal models, allowing for the study of different aspects of disease	Less well characterized, not widely used. Disease manifests primarily in the cecum	Prairie dogs are administered cefoxitin resulting in CDI [11]
Nonhuman primate model	Involves administration of CD toxins to rhesus macaques to study CDI	Closest to human physiology and disease progression	Expensive and ethically challenging	Rhesus macaques are administered CD toxins [12]
Piglet model	Involves the administration of CD spores to germ-free piglets	Piglets are naturally susceptible to CD	Not as widely used or accessible as smaller animals	Germ-free piglets are orally administered CD spores [13]

GI tract, specifically the colon, is primed for CD colonization, allowing researchers to inoculate a dose of either vegetative cells or CD spores orally, to achieve an infection that resembles that seen in humans. Here we use an established mouse model of CDI [23] and develop an optimal CD dosing strategy and treatment regime to be able to answer questions regarding treatment efficacy by reporting on several parameters of CDI (*see* **Notes 1** and **2**).

2 Materials

2.1 Mouse Strain, Stock Breed

1. Mice used in these studies consisted of both male and female C57BL/6 from The Jackson Laboratory, aged between 2 and 3 weeks and used at ages between 5 and 6 weeks old (*see* **Note 3** *regarding the choice of mouse strain*).
2. Sterile caging must be prepared to house each group of mice.
3. Mice are split into four different groups according to the University Animal Care housing guidelines (ex: up to three males and five females per cage). Groups include a negative vehicle control group, a treatment control group, a CD control group, and a CD + treatment group (Table 2).

Table 2
Description of groups and oral gavage inoculum for each day

	Treatment days	CD infection day
Negative/vehicle control	Treatment vehicle control	BHI media
Treatment control	Treatment	BHI media
CD control	Treatment vehicle control	CD culture
CD+ treatment	Treatment vehicle control	CD culture

4. If collecting stool daily (or only on specific days), new sterile cages must be prepared to transfer mice to on stool collection days (*see* **Note 4**).

2.2 Antibiotics

1. Mice will receive 0.4 mg/mL kanamycin, 0.035 mg/mL gentamicin, 850 U/mL colistin, 0.215 mg/mL metronidazole, and 0.045 mg/mL vancomycin in double distilled water per day.
2. 200 mL antibiotic drinking water should be prepared per cage, per day (for a total of 3 days).
3. Clindamycin (intraperitoneal injection).

   Mice will receive 10 mg/kg of clindamycin via intraperitoneal injection.

   Injections were performed using 26-gauge needle.

2.3 CD Strain and Culture Media

1. CD used in these studies is CD ribotype-027 (*see* **Note 5**).
2. Grown on cycloserine-cefoxitin fructose agar (CCFA) plates in anaerobic conditions.
3. CCFA plates may be made in-lab or pre-made plates may be purchased.
4. In-house ingredients for 1 L: 40 g proteose peptone No. 2, 5 g disodium phosphate, 1 g monopotassium phosphate, 6 g fructose, 20 g agar, 1 L ddH_2O, and 3 ml neutral red indicator (1% in ethanol), adjust pH to 7.6. Autoclave for 45 min at 121 °C. Add antibiotics (after cooling): 500 mg cycloserine and 16 mg cefoxitin. A single colony from the culture on the CCFA plate is inoculated into liquid brain–heart infusion media (BHI, BD Biosciences, USA).

2.4 Treatment Preparation

1. Treatment is prepared in appropriate aqueous solution for oral gavage at the desired dose.
2. In our model, we dose mice before CD inoculation, during, and post-infection daily out to study endpoint to account for our protein's 12-h half-life.

2.5 Euthanization and Tissue Storage

1. Surgical tools to perform euthanization.
2. Forceps.
3. Scissors.
4. 26-gauge needle and syringe for cardiac puncture.
5. 15 mL Conical tubes (1 per mouse, per organ) pre-filled with 5–10 mL of 10% phosphate-buffered formalin.
6. 1.5 mL Eppendorf tubes (one per mouse, per organ).
7. Printed rulers for measuring colonic length (one per mouse).

3 Methods

3.1 In Lab Preparation (Fig. 1)

1. Antibiotic water: Prepare 200 mL of antibiotic water per cage of mice, using sterile double distilled water. Add and dissolve: kanamycin 0.4 mg/mL, gentamicin 0.035 mg/mL, colistin 850 U/mL, metronidazole 0.215 mg/mL, and vancomycin 0.045 mg/mL.
2. Clindamycin: Prepare 2 mg/mL clindamycin in distilled water from stock.
3. CD culture: De-gas CCFA plates (in anaerobic chamber or bags/boxes with anaerobic pouches) overnight prior to CD culture.
4. CD ribotype-027 is cultured from frozen stock on CCFA plates in anaerobic conditions (chamber or bags/boxes) for 48 h.

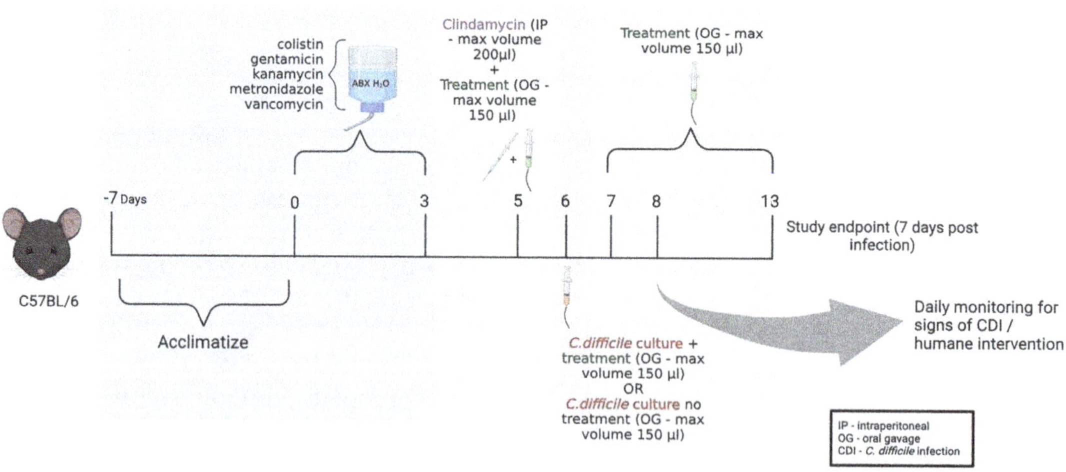

Fig. 1 Experimental timeline. Schematic diagram of our mouse model experimental timeline including days post-infection. Colored boxes highlight different arms of the study including pre-infection (blue), pre- and post-infection treatment (green), and CD infection (red)

One colony of CD from CCFA plates is transferred to 10 mL of BHI broth and grown for 48 h.

5. Prepare 1.0 McFarland of CD vegetative cells the day of infection by diluting CD liquid culture with BHI and measuring optical density using a biological density meter (e.g. DensiChek Plus).

3.2 Pre-infection (Days 0–7): Acclimation (See Note 3)

1. Upon arrival at the animal care facility, mice are transferred to sterile caging, with sterile water, and autoclaved chow.
2. Cage groupings may vary based on experimental groups and the sex of mice being used; e.g., females may be housed in groups of up to $n = 5$, while males should be housed in smaller groups ($n = 1$ or 2) and separated if fighting.
3. Allow mice to acclimatize for up to 7 days upon arrival at the university animal care facility.

3.3 Pre-infection (Days 1–3): Antibiotic Water

1. Following acclimatization, regular water from each cage is replaced with 200 mL of antibiotic water containing 0.4 mg/mL kanamycin, 0.035 mg/mL gentamicin, 850 U/mL colistin, 0.215 mg/mL metronidazole 0.215, and 0.045 mg/mL vancomycin as previously mentioned (*see* **Note 5**).
2. Option to collect "pre-antibiotic" fecal samples at this time point, on Day 0. To collect fecal pellets per cage, place all mice into a new sterile cage and use forceps to collect fecal pellets from old cage and place in Eppendorf tubes.
3. Repeat for the next 3 days, replacing the cage antibiotic water with freshly prepared antibiotic water.
4. Option to collect "post-antibiotic" fecal samples on Day 3. Transfer mice to a new sterile cage and collect stool from the old, empty cage.
5. On Day 3, replace antibiotic water with fresh, sterile water for the remainder of the study (*see* **Note 6**).

3.4 Pre-infection (Day 5): Clindamycin Injection and Pretreatment

1. On Day 5 mice are pretreated with the experimental therapeutic (or a vehicle control) via oral gavage (Table 2) using techniques in accordance with the university animal care SOP for oral gavage in small animals.
2. On the same day mice also receive an intraperitoneal injection of clindamycin, using techniques in accordance with the university animal care SOP for intraperitoneal injection in small animals.
3. Mice are injected with 10 mg/kg of clindamycin based on their weight on that day. Based on a 2 mg/mL preparation of clindamycin, examples of injection volumes for the following

weights are listed below. Collect "pretreatment" fecal samples at this time point, on Day 5.

21 g mouse—105 μL

20 g mouse—100 μL

19 g mouse—95 μL

18 g mouse—90 μL

17 g mouse—85 μL

16 g mouse—80/μL.

3.5 CD Inoculation (Day 6)

1. On Day 6 mice receive 100 μL of 1 McFarland of CD culture (or vehicle control) via oral gavage (Table 2) using techniques in accordance with the university animal care SOP for oral gavage in small animals (*see* **Note 5** *related to strain of CD to use for infections*).
2. Option to combine a follow-up dose of treatment with CD culture for a total of 150 μL (maximum oral gavage volume allowed in mice).
3. Collect "pre-infection" fecal samples at this time point, on Day 6.

3.6 Post-infection (Days 7–Endpoint): Daily Treatment

1. On Day 7 mice receive a follow-up post-treatment dose via oral gavage or vehicle control (Table 2), using techniques in accordance with the university animal care SOP for oral gavage in small animals.
2. Mice receive follow-up doses daily until humane or study endpoint.
3. Humane endpoint is defined based on University Animal Care Committee (UACC) approved animal use protocol but should include surrogate markers for CDI such as weight loss (up to 25% from pre-treatment), hunched posture, immobility, and bloody diarrhea.
4. Collect "post-treatment" fecal samples daily.

3.7 Postmortem/ Reporting

1. Since death is an outcome of CDI in mice, Kaplan–Meier survival curves are generated based on mortality rates throughout the experiment and are reported as a measure of treatment efficacy.
2. At study or humane endpoint, mice are euthanized in accordance with the University Animal Care Committee-approved animal use protocol. Euthanasia via cardiac puncture is ideal for blood collection.
3. At endpoints, the entire large intestine (rectum to and including cecum) should be harvested from each mouse, placed in a transparent dish on top of a printed ruler, and then length can be determined as a surrogate marker for colitis severity (Fig. 2).

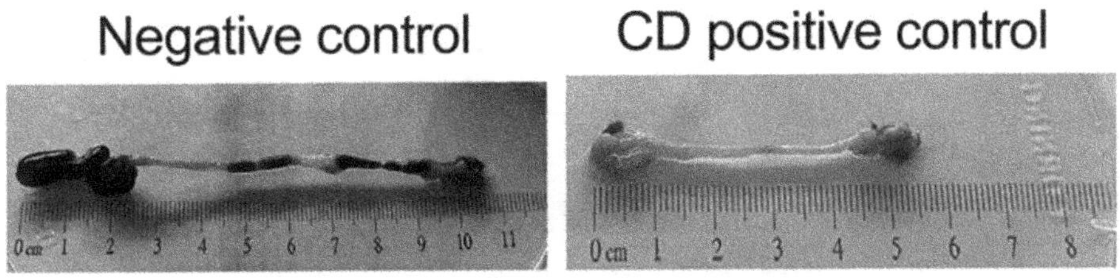

Fig. 2 Colonic length as a surrogate marker of CDI. Representative colons from a negative control mouse (left) and a CD positive control mouse (Right). Colonic length is measured using printed disposable rulers and recorded in a database for each colon. Colons from each group can be measured and total length can be averaged as a parameter for quantifying CDI. Shorter colon lengths indicate greater colitis severity.

4. A section of the distal colon should be stored (with rectum attached to easily identify the distal portion) and immediately stored in 10% phosphate-buffered formalin at room temperature. Tissues should remain in 10% phosphate-buffered formalin for 24 h to 2 weeks and then be transferred to 100% ethanol.
5. Histological processing and hematoxylin and eosin (H&E) staining: fixed tissues are dehydrated with ethanol, cleared with xylene and embedded in paraffin wax. 4 μm sections are transferred to slides and stained with H&E.
6. Immunohistochemistry may also be performed on fixed tissue to investigate myeloperoxidase (MPO, a marker of inflammation) or tight junction proteins.
7. Histology scoring: Our scoring system of histological images was adapted from Shelby et al., where images are scored from 0 to 3 for epithelial tissue damage, submucosal edema, and neutrophil infiltration, with zero being no damage and three indicating severe damage for a total maximum score of 9 [24] (*see* **Note** 7, Fig. 3).
8. A section of the distal colon may be stored alone in Eppendorf tubes at −80 °C for non-histological downstream analyses such as western blotting or ELISA.
9. Additional tissue such as the cecum, small intestine, liver, spleen, and kidney can also be harvested and stored in the same conditions for desired downstream analyses.
10. Stool collected throughout the experiment and postmortem is used for downstream analyses.

3.8 Analysis to Quantify Infection Outcomes

Cell toxicity assays using mouse stool: Fibroblast cells are susceptible to TcdA/TcdB-mediated disruption of the actin cytoskeleton causing morphological changes in cells. Incubating fibroblast cells with mouse stool supernatant can provide a measure of toxin activity in mouse stool. Specific methods for this are described below:

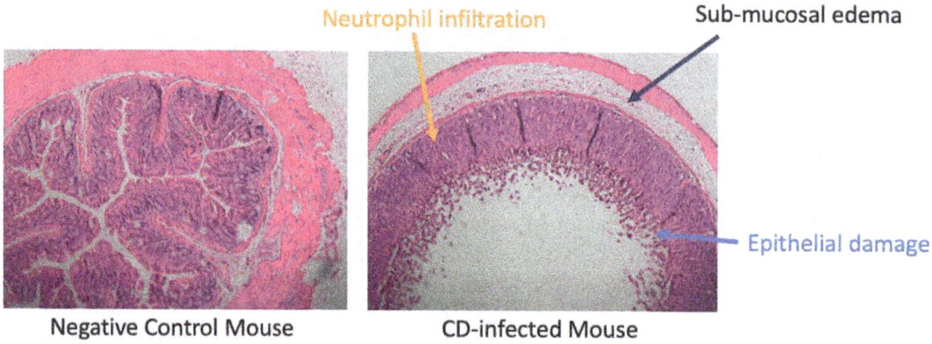

Parameter	Negative Control	CD Infected Mouse
Neutrophil Infiltration	0	1
Submucosal Edema	1	2
Epithelial Damage	0	3

Fig. 3 Infection histology: Example of distal colon histology scoring

1. NIH 3 T3 fibroblast cells are cultured in Dulbecco's Modified Eagle Medium (DMEM) supplemented with 10% fetal calf serum (FCS) in tissue-culture treated culture dishes at 37 °C with 5% CO_2.
2. Once cells have grown to ~70% confluency, cells are seeded into 24-well cell culture dishes containing 1 mL (total volume including cells) DMEM with 10% FCS and incubated at 37 ° C with 5% CO_2 until cells in wells reach ~70% confluency.
3. 10 μL of stool supernatant depending on toxin activity and concentration should be added to each well containing 490 μL of DMEM with 10% FCS for a final volume of 500 μL. Cells are placed back in the incubator 37 °C with 5% CO_2, and cells were monitored periodically (approximately every 30–60 min) for rounding (morphological changes caused by rearrangement of actin cytoskeleton). Stool supernatant from CD control mice should be used as a positive control, and experiment should be stopped when ~70–80% of cells in these wells have rounded.
4. To stop the experiment and fix cells, remove media from wells using transfer pipette (being careful to only touch the side of the wells, not the cells on the bottom). Then wash the cells once with PBS, apply 10% formalin-buffered phosphate to cells and leave for 15–20 min, remove formalin and wash cells with PBS, and then apply Giemsa stain to cells and leave overnight at room temperature on shaker.

5. Remove Giemsa stain from cells, and wash twice with PBS. Apply PBS to cells and image using inverted microscope.
6. TcdA/TcdB ELISAs may be used to quantify stool toxin concentrations.
7. MPO quantification: Myeloperoxidase, an inflammatory enzyme, may be quantified in stool to provide an indication of inflammation using spectrophotometric assays that measure MPO activity.

4 Notes

1. Here we provide an overview of the animal models that can be used in the development of novel therapies for *CD*. The models offer excellent insight into the pathogenesis of *CD* and monitor toxin activity (or neutralization). We have outlined specific and detailed protocols for assays to measure toxin as well as provided an outline of the different animal models and their benefits and setbacks. Despite the fact that animal models may not mimic the human condition precisely, animal models allow for more in-depth tissue analysis including tissue histology, evaluation of gross surrogate markers of disease including colon length, controlled collection of biospecimens during infection including of blood for cytokine and chemokine analysis, stool for microbiome or metabolome analysis, and tissue specimens for molecular or analytic interrogation [25, 26] all of which can be challenging, and resource intensive or near impossible to collect from patients.
2. As with any in vivo surrogate animal models, there are limitations that require consideration when designing experiments including the variability in disease presentation and/or outcome that may be present depending on the genetic background of the animals used. However, mice are a more clinically relevant model compared to other small animal models such as hamsters as the disease in mice clinically manifest to what is seen in humans where the animals develop diarrhea, and the pathogen primarily replicates and cause inflammation in the colon. This is different to what is observed in the guinea pig model, where the infection and inflammation occur in the cecum.
3. The most common murine background used is the C57BL/6 mouse model that was outlined above represents a very acute disease model that does not fully emulate the clinical disease progression seen in humans [27]. The model results in almost 90% murine mortality within the first 72 h of infection, a phenotype that is not typical in human beings [27]. There have been other models that have been developed to better

emulate the disease patterns seen in humans whereby investigators have changed the concentration of *CD* spores used to infect mice which have resulted in higher rates of survival in mice and resulting in prolonged colonization [28, 29]. But even in these models, the acuity of infection is much higher than observed in humans with mice losing almost 20% of their bodyweight in the first 48 h [29].

4. Note that for every day throughout the experiment, mice must be weighed, to monitor for changes in body weight to identify humane endpoints and as a surrogate marker for CDI.

   All handling of mice and cages should be performed in sterile biosafety level 2 hood.

5. In our experience using lower concentrations of CD results in lower mortality post-infection; however, it does result in greater variability in outcomes between experiments with some experiments resulting in little or no clinical signs of infection (diarrhea, hunched posture) in mice. Most of the animal models described and widely used in the field use one strain of CD, namely, the NAP1/Ribotype-027 strain, which may not be representative of what is observed in clinical infections as there have been significant changes in strain epidemiology [30, 31] in the last decade. So, it is suggested that the choice of strain that is most relevant be used for the model system.
6. It is important to note that it is difficult to monitor antibiotic water intake by mice, especially when co-housed; therefore, individual mice may intake varying amounts of water altering their antibiotic exposure. As a result, the subsequent clindamycin injection is imperative for ensuring adequate antibiotic-mediated disruption of the gastrointestinal microbiota.
7. Since histology scoring is prone to subjectivity, it is best practice that at least two independent researchers score the tissue in a blinded fashion and then the scores are shared with a third researcher to collate and ensure consistency of scoring.

References

1. Chang MCJ, Grieder FB (2024) The continued importance of animals in biomedical research. Lab Anim 53:295–297. https://doi.org/10.1038/s41684-024-01458-4
2. Hernandez BG, Vinithakumari AA, Sponseller B, Tangudu C, Mooyottu S (2020) Prevalence, colonization, epidemiology, and public health significance of Clostridioides difficile in companion animals. Front Vet Sci 7: 512551. https://doi.org/10.3389/fvets.2020.512551
3. Bartlett JG, Moon N, Chang TW, Taylor N, Onderdonk AB (1978) Role of Clostridium difficile in antibiotic-associated pseudomembranous colitis. Gastroenterology 75:778–782
4. Chen S, Sun C, Wang H, Wang J (2015) The role of rho GTPases in toxicity of Clostridium difficile toxins. Toxins (Basel) 7:5254–5267. https://doi.org/10.3390/toxins7124874
5. Britton RA, Young VB (2012) Interaction between the intestinal microbiota and host in Clostridium difficile colonization resistance.

Trends Microbiol 20:313–319. https://doi.org/10.1016/j.tim.2012.04.001

6. Chen X et al (2008) A mouse model of Clostridium difficile-associated disease. Gastroenterology 135:1984–1992. https://doi.org/10.1053/j.gastro.2008.09.002
7. Kaur S, Vaishnavi C, Ray P, Kochhar R, Prasad KK (2010) Effect of biotherapeutics on cyclosporin-induced Clostridium difficile infection in mice. J Gastroenterol Hepatol 25:832–838. https://doi.org/10.1111/j.1440-1746.2009.06135.x
8. Price AB, Larson HE, Crow J (1979) Morphology of experimental antibiotic-associated enterocolitis in the hamster: a model for human pseudomembranous colitis and antibiotic-associated diarrhoea. Gut 20:467–475. https://doi.org/10.1136/gut.20.6.467
9. Buckley AM, Spencer J, Candlish D, Irvine JJ, Douce GR (2011) Infection of hamsters with the UK Clostridium difficile ribotype 027 outbreak strain R20291. J Med Microbiol 60:1174–1180. https://doi.org/10.1099/jmm.0.028514-0
10. Oka K et al (2018) Establishment of an endogenous Clostridium difficile rat infection model and evaluation of the effects of clostridium butyricum MIYAIRI 588 probiotic strain. Front Microbiol 9:1264. https://doi.org/10.3389/fmicb.2018.01264
11. Muller EL, Pitt HA, George WL (1987) Prairie dog model for antimicrobial agent-induced Clostridium difficile diarrhea. Infect Immun 55:198–200. https://doi.org/10.1128/iai.55.1.198-200.1987
12. Arnon SS et al (1984) Rapid death of infant rhesus monkeys injected with Clostridium difficile toxins a and B: physiologic and pathologic basis. J Pediatr 104:34–40. https://doi.org/10.1016/s0022-3476(84)80585-5
13. Steele J, Feng H, Parry N, Tzipori S (2010) Piglet models of acute or chronic Clostridium difficile illness. J Infect Dis 201:428–434. https://doi.org/10.1086/649799
14. Lessa FC et al (2015) Burden of Clostridium difficile infection in the United States. N Engl J Med 372:825–834. https://doi.org/10.1056/NEJMoa1408913
15. Pereira JA, McGeer A, Tomovici A, Selmani A, Chit A (2020) Incidence and economic burden of Clostridioides difficile infection in Ontario: a retrospective population-based study. CMAJ Open 8:E16–E25. https://doi.org/10.9778/cmajo.20190018
16. McGlone SM et al (2012) The economic burden of Clostridium difficile. Clin Microbiol Infect 18:282–289. https://doi.org/10.1111/j.1469-0691.2011.03571.x
17. Feuerstadt P, Theriault N, Tillotson G (2023) The burden of CDI in the United States: a multifactorial challenge. BMC Infect Dis 23:132. https://doi.org/10.1186/s12879-023-08096-0
18. Louie TJ et al (2011) Fidaxomicin versus vancomycin for Clostridium difficile infection. N Engl J Med 364:422–431. https://doi.org/10.1056/NEJMoa0910812
19. Lowy I et al (2010) Treatment with monoclonal antibodies against Clostridium difficile toxins. N Engl J Med 362:197–205. https://doi.org/10.1056/NEJMoa0907635
20. Hvas CL et al (2019) Fecal microbiota transplantation is superior to Fidaxomicin for treatment of recurrent Clostridium difficile infection. Gastroenterology 156:1324–1332. https://doi.org/10.1053/j.gastro.2018.12.019
21. Cammarota G et al (2015) Randomised clinical trial: faecal microbiota transplantation by colonoscopy vs. vancomycin for the treatment of recurrent Clostridium difficile infection. Aliment Pharmacol Ther 41:835–843. https://doi.org/10.1111/apt.13144
22. Kelly CP (2012) Can we identify patients at high risk of recurrent Clostridium difficile infection? Clin Microbiol Infect 18(Suppl 6):21–27. https://doi.org/10.1111/1469-0691.12046
23. Douchant K et al (2024) Defined microbial communities and their soluble products protect mice from Clostridioides difficile infection. Commun Biol 7:135. https://doi.org/10.1038/s42003-024-05778-6
24. Shelby RD et al (2020) Development of a standardized scoring system to assess a murine model of Clostridium difficile colitis. J Investig Surg 33:887–895. https://doi.org/10.1080/08941939.2019.1571129
25. Hernandez Del Pino RE, Barbero AM, Espanol LA, Morro LS, Pasquinelli V (2021) The adaptive immune response to Clostridioides difficile: a tricky balance between immunoprotection and immunopathogenesis. J Leukoc Biol 109:195–210. https://doi.org/10.1002/JLB.4VMR0720-201R
26. Biazzo M, Allegra M, Deidda G (2022) Clostridioides difficile and neurological disorders: new perspectives. Front Neurosci 16:946601. https://doi.org/10.3389/fnins.2022.946601
27. Czepiel J et al (2019) Clostridium difficile infection: review. Eur J Clin Microbiol Infect Dis 38:1211–1221. https://doi.org/10.1007/s10096-019-03539-6

28. Mizrahi A et al (2024) A mouse model of mild Clostridioides difficile infection for the characterization of natural immune responses. Microorganisms 12. https://doi.org/10.3390/microorganisms12101933
29. Sun X et al (2011) Mouse relapse model of Clostridium difficile infection. Infect Immun 79:2856–2864. https://doi.org/10.1128/IAI.01336-10
30. McDonald LC et al (2005) An epidemic, toxin gene-variant strain of Clostridium difficile. N Engl J Med 353:2433–2441. https://doi.org/10.1056/NEJMoa051590
31. Katz KC et al (2018) The evolving epidemiology of Clostridium difficile infection in Canadian hospitals during a postepidemic period (2009-2015). CMAJ 190:E758–E765. https://doi.org/10.1503/cmaj.180013

Chapter 10

Purification of Type IV and Type VI Lipoteichoic Acids from Gram-Positive Mucosal Pathogens via a Two-Step Hot Water–Phenol/Triethylammonium Acetate Extraction Method

Jolene M. Garber, Harald Nothaft, Christine M. Szymanski, and Cory Q. Wenzel

Abstract

Type IV and Type VI lipoteichoic acids (LTAs) are found on the surface of Gram-positive mucosal pathogens. There is interest to purify, characterize, and evaluate these compounds as vaccine candidates. In traditional hot water–phenol extractions, Type IV and VI LTAs are found almost exclusively in the phenol fraction, and, due to the lengthy removal process, prolonged exposure to phenol can degrade them. Here we describe an efficient and time-saving new method to rapidly separate these LTAs from phenol through the addition of triethylammonium acetate (TEAA), and we present a continuous flow dialysis arrangement to rapidly remove excess phenol. We demonstrate the efficacy of this protocol through silver staining and immunoblotting of Type IV LTA from *Streptococcus pneumoniae* and Type VI LTA from *Clostridium perfringens*.

Key words Lipoteichoic acids, Gram–positive, Extraction, Purification

1 Introduction

Gram-positive mucosal pathogens can readily colonize their hosts and proliferate using their ability to degrade and assimilate mucosal carbohydrates and invade host tissues. For instance, both the human nasopharyngeal pathogen *Streptococcus pneumoniae* and the poultry gastrointestinal pathogen *Clostridium perfringens* possess large repertoires of extracellular glycoside hydrolases for the degradation of mucin glycans that are essential for their fitness [1, 2]. These infections are a growing concern as the rates of antimicrobial resistance (AMR) are increasing worldwide, with *S. pneumoniae* among the top six AMR pathogens causing death in 2019 [3]. Not only does the rise in AMR limit treatment options

Inka Brockhausen (ed.), *Dynamics of Bacteria-Mucus Interactions*, Methods in Molecular Biology, vol. 2942, https://doi.org/10.1007/978-1-0716-4627-4_10,

for human patients, but it has also led to policy changes for antimicrobial use within the livestock industry. While this is a necessary measure to save human lives, it has had economic consequences in the agricultural sector through the increase in certain livestock diseases that were once prevented by prophylactic antimicrobials. This is particularly true for necrotic enteritis caused by *C. perfringens* in poultry [4]. Thus, there is a need to develop alternative strategies for preventing these infections in both humans and animals such as the development of glycoconjugate vaccines.

Current glycoconjugate vaccines against Gram-positive microbes target the highly variable capsular polysaccharides (CPS), which have limitations since alternate CPS serotypes can readily emerge—as has been observed for *S. pneumoniae* after vaccination with the Prevnar 13® vaccine [5]. One approach to handling this challenge has been to improve the throughput of glycoconjugate synthesis through the use of engineered bacterial strains for production of bioconjugates as an alternative to chemical conjugates, allowing for greater ease of producing multivalent *S. pneumoniae* vaccines [6, 7]. However, alternative strategies are to consider surface polysaccharides other than CPS. Another conserved surface glycan found in Gram-positive cell envelopes is lipoteichoic acid (LTA), typically consisting of various glycosyl phosphate polymers linked to a diacylglycerol-based lipid anchor [8–10]. LTAs have been further characterized into different types based on their structural properties. Differences in LTA types can impact how they are purified from bacterial cultures for assessing immunogenicity and potential as vaccine candidates. Both the *S. pneumoniae* Type IV [10] and the recently described *C. perfringens* Type VI [11] LTAs are unique due to the presence of phosphate modifications as side chain residues. Type IV LTAs contain phosphocholine side chain modifications whereas Type VI LTAs are modified with phosphoethanolamine and, to a lesser degree, phosphoglycerol [11]. Interestingly, both these LTAs are unexpectedly found in the phenol phase following a hot water–phenol extraction [11, 12], which presents a challenge for rapid purification due to the need to rapidly remove phenol to prevent glycan degradation. In contrast to the traditional phenol extraction technique, we describe a new method in this protocol where triethylammonium acetate (TEAA) solution is added to the phenol fraction to transition Type IV and Type VI LTAs into the aqueous phase. This, in combination with the rapid dialysis arrangement described, allows for faster purification of these glycans and reduces the contact time with phenol to minimize phenol-mediated degradation.

2 Materials

Prepare all solutions using ultrapure water (18 Ω at 25 °C). Prepare and store all solutions at room temperature unless otherwise indicated, and diligently follow all waste disposal regulations when disposing of waste materials.

2.1 Required Equipment/Supplies

1. Thermal shaking water bath, magnetic stir bars, a stir plate that can be placed in a 37 °C warm room or incubator.
2. Phenol-resistant gloves (consult manufacturer and/or your institution's health and safety guidelines).
3. 3.5 kDa molecular weight cutoff (MWCO) dialysis tubing.
4. Chemical fume hood with water faucet.
5. Dialysis vessel (see Subheading 3).
6. Lyophilizer.
7. Glass scintillation vials.
8. Kimwipes or similar lint-free porous sheets.
9. Parafilm.
10. Insulated nonporous container or bucket (for the dry ice and ethanol bath).
11. Centrifuge, centrifuge bottles that can be dedicated to phenol extractions (*see* **Note 1**).
12. 1 L overnight culture of the organism of interest grown under appropriate conditions.

2.2 Phenol–Hot Water Extraction

1. Liquefied phenol (~90% w/v) (for preparation from crystalline phenol, *see* **Note 2**).
2. 125 mM saline: weigh 7.3 g of sodium chloride (NaCl), dissolve in 900 mL of water, and adjust the final volume to 1 L.

2.3 Triethylammonium Acetate (TEAA) Extraction and Enzymatic Digestion

1. 1 M triethylammonium acetate solution: Santa Cruz Biotechnology (Dallas, TX) Catalogue No. SC-505910 or similar. Alternatively, this can be made by adding 140 mL of triethylamine (Thermo Fisher) into 800 mL of water and then slowly adding glacial acetic acid (Thermo Fisher Scientific) over several hours with stirring until a pH of 7.0 is achieved (this will require ~60 mL of glacial acetic acid).
2. 20 mg/mL DNAse I: weigh 20 mg of DNAse I (Thermo Fisher Scientific), and resuspend in 1 mL of water. Store at −20 °C, and thaw at room temperature when needed (*see* **Note 3**).
3. 20 mg/mL RNAse A: weigh 20 mg of RNAse A (Thermo Fisher Scientific), and resuspend in 1 mL of water. Store at

−20 °C, and thaw at room temperature when needed (*see* **Note 3**).

4. 20 mg/mL proteinase K: weigh 20 mg of proteinase K (Thermo Fisher Scientific), and resuspend in 1 mL of water. Store at −20 °C, and thaw at room temperature when needed (*see* **Note 3**).
5. 1 M $MgCl_2$: weigh 20.16 g of magnesium chloride hexahydrate ($MgCl_2 \cdot 6H_2O$) (Thermo Fisher Scientific), dissolve in 70 mL of water and adjust to 100 mL.

3 Methods

Carry out all procedures at room temperature unless otherwise specified.

3.1 Bacterial Growth

1. Grow a 1 L liquid culture of the bacterium of interest in the appropriate medium (*see* **Note 4**) and pellet by centrifugation (aim for a pellet with a wet weight of at least ~1.5 g). Alternatively, cells can be grown on agar medium and carefully collected into PBS. Wash the cell pellet with approximately 20 mL PBS or whatever minimum volume is needed to obtain a uniform suspension (*see* **Note 5**).

3.2 Phenol–Hot Water Extraction

1. Place a thermal shaking water bath in a fume hood, install clamps that will hold centrifuge bottles snugly, fill with deionized water, and preheat to 70 °C.
2. Place 20 mL of liquefied phenol into a 50 mL conical or centrifuge bottle, tighten the lid well, and place in the 70 °C shaking water bath to preheat.
3. Resuspend the washed cell pellet in 40 mL of ultrapure water, place the suspension into a 125 mL Erlenmeyer flask, and subject to a 20 min liquid run in an autoclave. Allow the sample to cool (*see* **Note 6**), and then separate the sample by centrifugation and decant the supernatant (*see* **Note 7**).
4. Resuspend the cell pellet in 20 mL of 125 mM saline. In the fume hood, remove the preheated phenol from the water bath, transfer the cell suspension into the phenol, replace the lid on the centrifuge bottle, and tighten well.
5. Shake the centrifugation bottle containing the cell suspension/phenol mixture vigorously by hand for 15 s, place the bottle back in a clamp in the water bath, and turn on the shaking water bath to the maximum speed possible without the water spilling out of the bath; aim for at least 150 rpm (*see* **Note 8**).
6. Continue to incubate the sample with shaking for 1 h, stopping briefly every 10 min to remove the samples from the bath,

carefully loosen lids to vent samples, retighten lids, shake vigorously by hand, and then replace in the water bath (as in Subheading 3.2, **step 5**).

7. Remove the sample from the water bath and allow to sit at room temperature for 10 min. Carefully loosen the lid of the centrifuge bottle slowly to minimize any risk of spray due to any remaining pressure in the bottle, and then place the sample on ice and allow to cool thoroughly (*see* **Note 9**).
8. While the sample is cooling, determine the length of 3.5 kDa MWCO dialysis tubing that will be needed to accommodate *at least* double the volume of the phenol phase, as the volume will increase significantly during dialysis (depending on the dimensions of your dialysis vessel, this may require splitting the phenol phase into more than one length of dialysis tubing). Trim the piece(s) of dialysis tubing required (*see* **Note 10**), and soak in a beaker with tap water for 30 min.
9. Once the sample is cool, tighten the lid on the centrifuge bottle, and subject the sample to centrifugation (13,000× *g*, 30 min) to separate the aqueous, phenol, and solid phases. There may be a white layer of denatured protein at the interphase between the (upper) aqueous phase and the (lower) phenol phase, as well as a pellet of cellular debris (that may be quite large, depending on the bacteria used).
10. Working in the fume hood, carefully remove the (lower) phenol phase by pipette (*see* **Note 11**).
11. Allow the phenol phase to cool to room temperature. While the sample is cooling, prepare the dialysis setup according to one of the two options in Fig. 1, rinse the dialysis tubing against running tap water (to remove residual glycerol), and tie a knot (*see* **Note 10**) and/or place a dialysis clamp on one end of each piece of dialysis tubing.

3.3 TEAA Extraction and Enzymatic Digestion

1. Once the phenol fraction has cooled, add 20 mL of TEAA solution (1 M) and shake vigorously for 15 s. Rest for 15 s and repeat the 15 s vigorous shaking.
2. Separate the TEAA and phenol phases by centrifugation as in 3.2 Item 9, and then carefully remove the (upper) TEAA phase via pipette.
3. Transfer the sample via pipette, 1 mL at a time, into the presoaked and rinsed 3.5 kDa-MWCO dialysis tubing (*see* **Note 12**). Carefully work the air bubbles out of the tubing, place another dialysis clamp and knot on the unclamped end of the tubing, and place into the receptacle of your dialysis setup, ensuring that the flow rate of tap water is as slow as possible while flowing continuously (i.e., not dribbling or spattering).

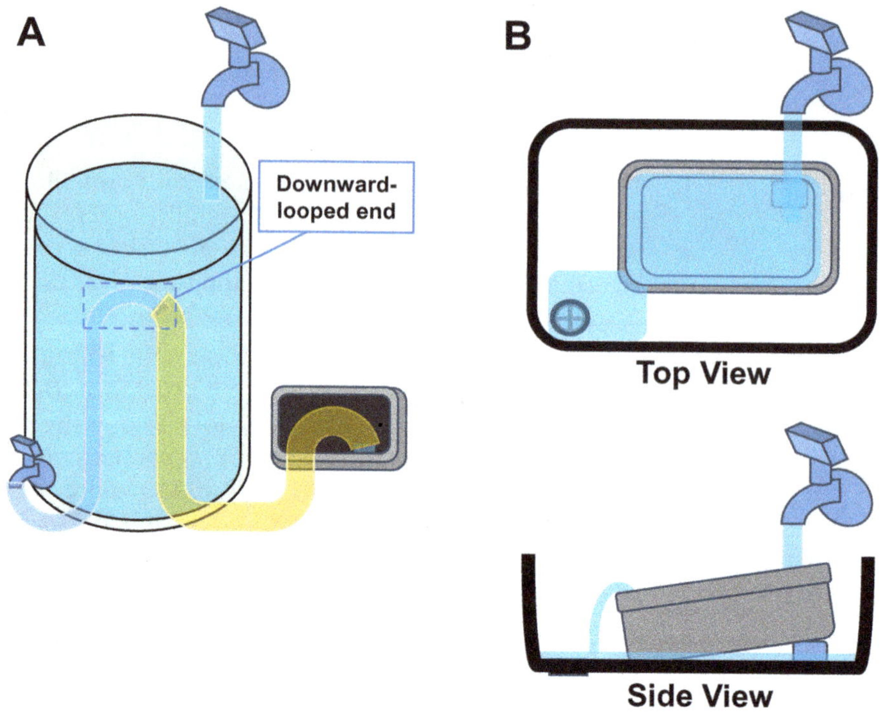

Fig. 1 Continuous-flow dialysis setup for rapid removal of phenol from extraction samples. (**a**) Glass drink-dispenser-based dialysis setup. (**b**) Basin-based dialysis setup. Dialysis should be performed in (i) a fume hood with a water faucet and sink or (ii) a well-ventilated room with a sink for which access can be restricted for 3–5 days; for the former, the setup in (**a**) is preferred, due to most fume hoods only having small cup sinks, while the setup in (**b**) is preferable if a suitable low-use room is available. The setup in (**a**) involves fastening a length of small-diameter clear PVC tubing (we use 5/16″ inner diameter, 7/16″ outer diameter) to the spout at the bottom of a glass drink-dispenser container (using an appropriate glue, cinching with a zip-tie, etc.). This tubing (blue in diagram) is then run alongside the outside of the container and looped downwards near the top of the container and fastened in-place to the container via clear packing tape. The height of the top of this loop will determine the steady-state volume of the water inside the container during dialysis. A second length of wider-diameter tubing (yellow in diagram) is then placed over the downward-looped end of the small-diameter tubing (as indicated by the boxed area in the diagram) and should be long enough to channel water flow from the small-diameter tubing into the fume hood cup sink when the container is placed under the fume hood faucet. **IMPORTANT:** *the wider-diameter tubing must be wide enough to prevent an airlock from forming between the two lengths of tubing.* If an airlock forms, then siphoning action will empty the container to the level of the bottom spout. This assembly is placed under the faucet in the fume hood, filled with water, and then the flow of water from the faucet adjusted to a slow steady stream of water. As long as the water is not flowing faster than the maximum flow capacity of the smaller diameter tube, this arrangement will maintain the water level at the height of the upper curved section of that tubing while the continuous flow of water ensures constant exchange of water to increase the rate at which the phenol is dialyzed away. **IMPORTANT**: *prepare this set-up a few hours before it will be left unattended so that it can be monitored for airlocks within the small-diameter tubing.* Shortly after filling and early during constant-flow from the faucet, air bubbles can be trapped in the tubing but can be coaxed out by either tapping or gently shaking the smaller-diameter tubing at the lower loop that forms by its attachment point to the container spout. Once the system has been flowing without issues for 30 min, samples can be loaded into the dialysis tubing and can be placed into the container. The setup in (**b**) involves placing a large basin in a sink such that one corner is directly below the faucet, with a rubber stopper placed under this corner of the basin. This results in a raised corner where fresh water enters the basin, and the resultant slope of the basin results in water flowing out of the opposing corner of the basin. The basin is then filled with water, followed by adjustment of the water flow to a slow continuous stream, at which point samples in the dialysis tubing can be placed into the basin

4. Dialyze this fraction until phenol can no longer be detected through smell when the tubing is ~2 cm from the nose. In our hands a single overnight dialysis is sufficient, but this is dependent on the dimensions of the dialysis container, the size and surface area of the tubing, flow rate of the water, etc. so a second day of dialysis may be necessary.
5. Carefully remove the clamp from one end of the dialysis tubing (*see* **Note 13**), and transfer the contents to an appropriately sized container for the volume along with a suitably sized magnetic stir bar. Add DNAse I and RNAse A to a final concentration of 10 μg/mL each and $MgCl_2$ to a final concentration of 10 mM (*see* **Note 14**), and place on a stir plate at 37 °C for 4 h.
6. Add proteinase K to a final concentration of 15 μg/mL and incubate at 55 °C–65 °C overnight in a shaking water bath at 100 rpm.
7. Prepare dialysis tubing and the dialysis setup as in Subheading 3.2, **item 8** and **11**. Transfer the sample to the dialysis tubing, and dialyze overnight.

3.4 Lyophilization and Preparation of Standardized Aliquots

1. Transfer the dialyzed and enzymatically treated sample to an ethanol-resistant bottle or vial that holds at least double the volume of sample.
2. In a fume hood, prepare a dry ice/ethanol bath by adding enough ethanol to cover the dry ice pellets up to a volume of 250 mL. The ethanol will bubble vigorously when first added, due to CO_2 release as the dry ice cools down the ethanol.
3. Once the bubbling of the ethanol/dry ice mix is reduced to a slow stream of bubbles under the surface and the ethanol is slightly viscous, place the vial/tube containing the sample into the bath, on an angle to maximize the surface area of the inner wall being touched by the sample, and turn the vial/tube constantly until the sample is completely frozen in a shell along the walls of the tube, which both maximize surface area to reduce the lyophilization time and prevent the sample from rupturing the vial/tube, even if it is glass (*see* **Note 15**).
4. Remove the lid on the vial/tube, wipe around the rim with a paper towel to remove any moisture on the edge, then place a piece of kimwipe over the opening of the vial/tube, and secure with parafilm by wrapping at least two times around the outside wall, resulting in a vial/tube with a porous kimwipe "filter" over the top of the opening.
5. Lyophilize the sample according to the lyophilizer manufacturer's instructions, and remove once the sample is fully dehydrated (typically overnight to 2 days, depending on sample volume, surface area, and vacuum strength of the lyophilizer).

6. Resuspend the lyophilized sample in 8–10 mL of water by stirring with an appropriately sized magnetic stir bar on a stir plate at room temperature. While the sample is resuspending, pre-weigh a 20 mL glass scintillation vial with lid on an analytical balance, and note the mass. Add the resuspended sample (which may not fully dissolve, but should be fairly homogeneous), and freeze the sample in the scintillation vial using the dry ice/ethanol bath method described in Subheading 3.4, **item 2** and **3**.
7. Prepare the vial for lyophilization as in Subheading 3.4, **item 4** and lyophilize, as per Subheading 3.4, **item 5**.
8. Secure the lid to the scintillation vial and weigh the vial including the sample, and determine the mg of LTA by subtracting the mass of the empty vial (taken in Subheading 3.4, **step 4**).
9. Add MilliQ water to the sample to generate a 2 mg/mL solution, resuspend with stirring as in Subheading 3.4, **item 6**, until homogeneous (it will likely not fully dissolve), and partition into 500 μL aliquots in 2 mL microfuge tubes and freeze at −20 °C.
10. These aliquots can either be stored at −20 °C until needed as a solution, or they can be lyophilized to yield 1 mg aliquots that can be stored at room temperature (*see* **Note 16**). To lyophilize these samples, aliquot 500 μL of the 2 mg/mL solution by pipetting into 2 mL microfuge tubes and freeze on a slight angle at −80 °C so that the liquid runs up the side of the tube as much as possible but does not enter the lid. While waiting for the samples to freeze, carefully cut off lids from empty 2 mL microfuge tubes, and, with the removed lids placed on a firm surface, carefully poke a few holes into them with a 19-gauge needle. Immediately before lyophilizing, open the frozen sample tubes and top them with the perforated lids and place in the lyophilizer.
11. Once the sample is lyophilized completely, tap the perforated lid gently to dislodge any sample that may have ended up on the lid, remove the lid, cap the sample tube with its attached lid, and seal the tubes with parafilm. This will yield individual 1 mg aliquots. Perforated lids can be cleaned by soaking overnight in ethanol and rinsing thoroughly with distilled water to reuse in future preps. These samples can be sent for structural analysis or examined in silver-stained gels and western blots (Fig. 2).

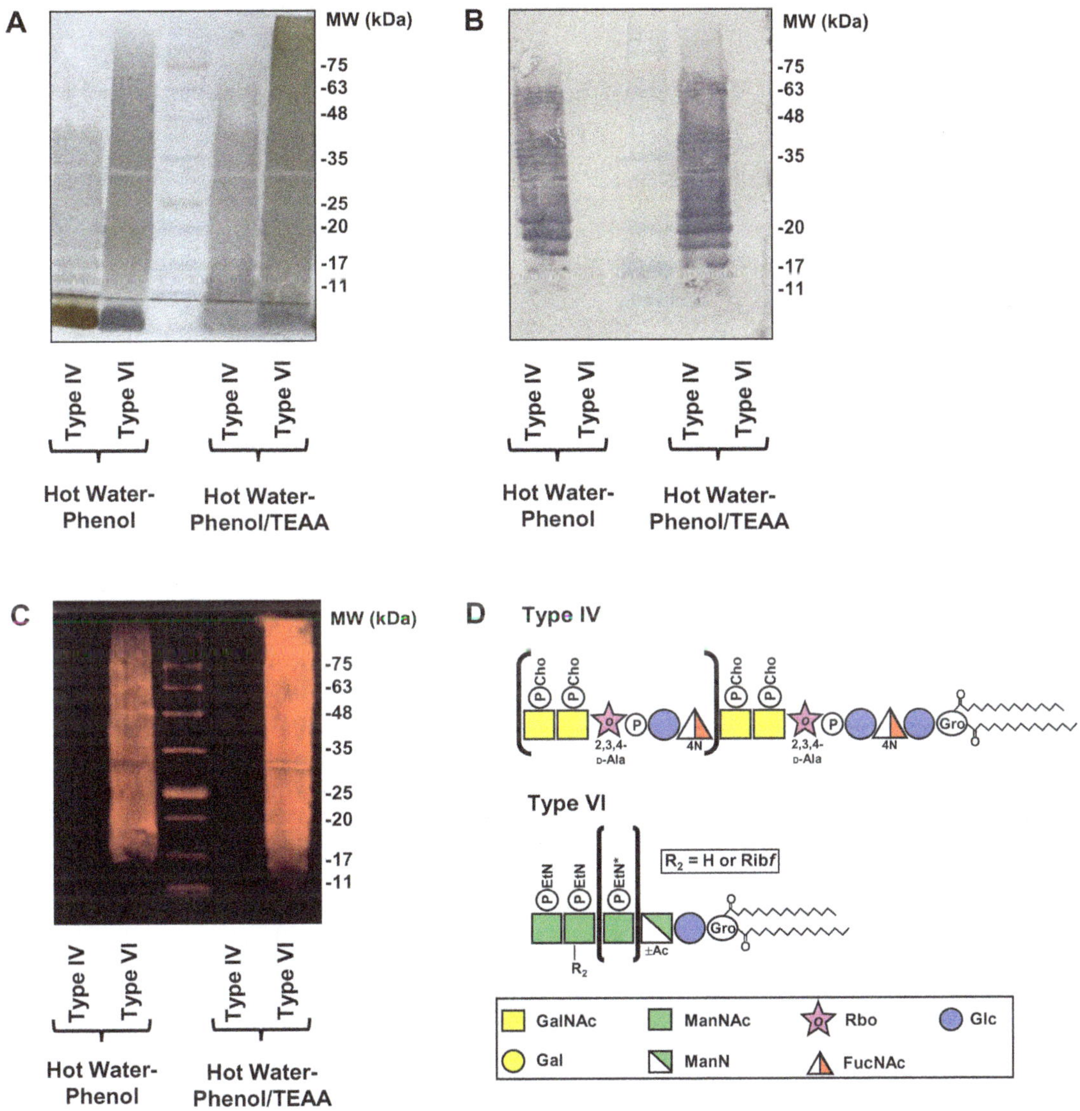

Fig. 2 Visualization of specific Type IV and VI lipoteichoic acids (LTAs) purified via phenol/TEAA extraction. (**a**) Silver stain comparing Type IV LTA from *S. pneumoniae* ATCC 6308, serotype 8 and Type VI LTA from *C. perfringens* HN13 [13]. Similar LTAs were isolated via the traditional hot water–phenol extraction (after rapid dialysis) and our new hot water–phenol/TEAA extraction methods. (**b**) Anti-phosphocholine (PCho) immunoblot to detect Type IV PCho-containing *S. pneumoniae* LTA [14], using mouse IgA TEPC 15 (Sigma Aldrich) as primary antibody (1:1000 dilution) and alkaline phosphatase-conjugated goat anti-mouse-IgA as the secondary antibody (1:1000 dilution) and visualized colorimetrically by development with NBT/BCIP. (**c**) Anti-Type VI *C. perfringens* LTA immunoblot using rabbit anti-*C. perfringens* HN13 LTA antiserum (as described in [11]). Purified LTA was suspended to 2 mg/mL as described above. Then 100 μL of this suspension was combined with 66.6 μL of 4× Laemmli buffer [15] and boiled for 20 min. After pelleting insoluble material by centrifugation, 30 μL of each sample were loaded into 12.5% separating/5% stacking SDS-PAGE gels and run at 180 V for 1 h. For (**a**), a modified version of the protocol by Tsai and Frasch [16] was used. Briefly, the gel was placed in a rocking glass tray of fixing solution (40% ethanol, 5% acetic acid in mqH_2O) for 2 h, after which it was placed in 50 mL of oxidizing solution (0.7% periodic acid in fixing solution) for 15 min, and washed 3 × 15 min with water. The gel was then placed in a freshly prepared pre-stain solution (60 mL water, 14 mL 0.1 N NaOH, and 1 mL 28% NH_4OH] (*see* **Note 17**) for 10 min and then into

4 Notes

1. The smell of phenol permeates the centrifuge bottles used for this method and cannot be fully removed; therefore, it is recommended to use centrifuge bottles that will be dedicated solely for use in phenol extractions.
2. Pure phenol is crystalline at room temperature and has a melting temperature of ~42 °C. To prepare liquefied phenol, add 55 mL of ultrapure water to a 500 g bottle of crystalline phenol, seal tightly, gently warm the 500 g bottle of phenol in a 45–50 °C water bath *in a fume hood*, gently swirling contents once the crystals have melted/dissolved until a homogeneous mixture has been obtained, then remove from heat, and allow to cool to room temperature.
3. Do not use commercially prepared DNAse, RNAse, or proteinase K solutions, as they typically contain glycerol which can interfere with downstream analytical work.
4. Care should be taken in selecting a medium that will not contaminate the sample with medium-derived glycans, especially if the LTA is being prepared for structural or composition analysis (i.e., nuclear magnetic resonance, mass spectrometry, gas chromatography, etc.); for many bacteria, brain–heart infusion medium works well, and our group has had success with PGY media (30 g proteose peptone No. 3, 20 g D-glucose, 10 g yeast extract per L) for both *C. perfringens* and *S. pneumoniae*.
5. To ensure efficient extraction, it is important that cell pellets be resuspended to a homogenous consistency at all steps. To avoid cell clumps, we have found it works well to initially only add 5 mL of the resuspension liquid to the pellet, dislodge the

Fig. 2 (continued) freshly prepared silver staining solution (57.5 mL water, 14 mL 0.1 N NaOH, 1 mL 28% NH_4OH, and 2.5 mL 20% silver nitrate [added dropwise with stirring]) (*see* **Note 18)** for 10 min and washed 3 × 10 min with water. Finally, the gel was placed in freshly prepared developer solution (3% sodium carbonate; 1.8% formaldehyde) in water which was exchanged frequently until bands appeared, and then development was stopped with 5% acetic acid. For the immunoblots in (**b**) and (**c**), gels were transferred in a semidry transfer cell to 0.2 μM nitrocellulose using a modified transfer buffer (25 mM Tris, 192 mM glycine, 0.1% SDS, pH 8.3 in 20% methanol) at 25 V for 1 h and blocked overnight at 4 °C in Intercept™ blocking buffer (LI-COR). Antibody incubation steps were performed with antibodies diluted in 50% Intercept/50% PBST (PBS with 0.05% Tween 20). (**d**) Schematic illustration using the Symbol Nomenclature for Glycans of Type IV and Type VI LTA (adapted from [17]). GalNAc = N-acetylgalactosamine, Gal = galactose, ManNAc = N-acetylmannosamine; ManN = mannosamine; Rbo = ribitol; FucNAc = N-acetylfucosamine; Glc = glucose; P = phosphate; Cho = choline; Gro = glycerol; EtN = ethanolamine, Rib*f* = ribofuranose, H = hydrogen; D-Ala = D-alanine; 4 N = 4-amino; ±Ac indicates that residue can be either ManNAc or ManN; *indicates that a minor percentage of residues can be substituted with PGro in place of PEtN pellet from the side of the tube with the side of a pipette tip, and then pipette up and down until the cell-liquid suspension becomes homogeneous. If the suspension is too viscous with only 5 mL, adjust this volume by adding 1 mL at a time until the suspension is loose enough to be mixed to homogeneity. Once a homogeneous suspension is obtained, the remaining volume of resuspension liquid can be added and the sample swirled well until a uniform suspension is obtained. If transferring between vessels, we recommend reserving a portion of the final resuspension liquid volume to rinse the first vessel after transferring the majority of the cell suspension to ensure complete transfer of most cells.

6. To save time, the sample can be rapidly cooled by incubating the flask in an ice water bath.
7. While the supernatant from Subheading 3.2, **item 3** does not typically contain LTA, other glycans and/or glycolipids (e.g., capsule) may be present in this fraction, so you may wish to save it for additional analyses.
8. When determining the maximum shaking speed that can be achieved without splashing water out of the water bath, splashing at higher speeds can often be eliminated by adjusting the volume of water up or down a small amount rather than by slowing the shaking speed.
9. Loosening the lid of the centrifuge bottle prior to chilling on ice is essential; if this is not done, it will be impossible to loosen the lid once the sample has been chilled.
10. We like to include an additional 20 cm of the tubing to allow for a knot to be tied outside of each dialysis clip, which acts as a backup measure to protect from sample loss should a clip fail and release.
11. The LTA from Gram-positive bacteria with Type IV or VI LTA should not partition to the aqueous phase. However, other LTA types may partition to the aqueous phase.
12. It is important that the dialysis tubing does not dry out when filling, so we recommend keeping a container of tap water nearby into which we will periodically dip the tube into while pinching off the open end during the filling process.
13. We have found it works well to hold the tube vertically so that most of the sample collects at one end of the tube and squeeze any remaining liquid down into it. If enough excess tubing was left, there should be a span of empty flattened tubing between the sample and the clamp. Carefully pinch above the sample and cut the tube in the flattened area below the clamp. Place the open end of the dialysis tube in the new vessel, and squeeze out the sample.

14. $MgCl_2$ is added due to DNAse I requiring Mg^{2+} for activity.
15. As the sample freezes, it will make a crackling sound and start to crack in a way that appears like shattered glass, but this is the sample cracking as the thin frozen layer expands and not the vial that it is contained within (as long as you are constantly rotating the sample in the bath since leaving the vial static in the dry ice/ethanol bath runs the risk of the sample freezing as a thick solid layer that may break or distort the vial/tube due to expansion and require more time to lyophilize due to decreased surface area).
16. Lyophilized glycans and glycolipids, such as LTA, are typically fluffy powders that are difficult to weigh out when in small quantities due to their tendency to retain static charge. By weighing a large preparation, dissolving the entire material to a specific concentration (e.g., 2 mg/mL) and then aliquoting the sample to defined quantities and lyophilizing, one can obtain small aliquots of sample with greater accuracy than by trying to weigh small amounts of sample on an analytical scale.
17. The addition of this step is to prevent formation of a darkened silver "film" appearing on the surface of the gel during development.
18. It is essential that the reagents for the silver staining solution are added in the order listed, that the NH_4OH is not too old (as NH_4OH solution loses potency over time due to evaporation of the volatile NH_4^+ contained within), and that the 20% silver nitrate is made fresh prior to preparing the staining solution.

Acknowledgments

J.M.G. received support from a fellowship from the Alberta Innovates Industry r&D Associates Program.

Conflict of Interest C.Q.W., J.M.G., H.N., and C.M.S. are affiliated with VaxAlta Inc.

References

1. Mathew BJ, Gupta P, Naaz T et al (2023) Role of *Streptococcus pneumoniae* extracellular glycosidases in immune evasion. Front Cell Infect Microbiol 13:1109449. https://doi.org/10.3389/fcimb.2023.1109449
2. Low KE, Smith SP, Abbott DW, Boraston AB (2021) The glycoconjugate-degrading enzymes of *Clostridium perfringens*: tailored catalysts for breaching the intestinal mucus barrier. Glycobiology 31(6):681–690. https://doi.org/10.1093/glycob/cwaa050
3. Antimicrobial Resistance Collaborators (2022) Global burden of bacterial antimicrobial resistance in 2019: a systematic analysis. Lancet 399(10325):629–655. https://doi.org/10.1016/s0140-6736(21)02724-0

4. Fathima S, Hakeem WGA, Shanmugasundaram R, Selvaraj RK (2022) Necrotic enteritis in broiler chickens: a review on the pathogen, pathogenesis, and prevention. Microorganisms 10(10). https://doi.org/10.3390/microorganisms10101958
5. Lo SW, Gladstone RA, van Tonder AJ et al (2019) Pneumococcal lineages associated with serotype replacement and antibiotic resistance in childhood invasive pneumococcal disease in the post-PCV13 era: an international whole-genome sequencing study. Lancet Infect Dis 19(7):759–769. https://doi.org/10.1016/s1473-3099(19)30297-x
6. Harding CM, Nasr MA, Scott NE et al (2019) A platform for glycoengineering a polyvalent pneumococcal bioconjugate vaccine using *E. coli* as a host. Nat Commun 10(1):891. https://doi.org/10.1038/s41467-019-08869-9
7. Price NL, Goyette-Desjardins G, Nothaft H et al (2016) Glycoengineered outer membrane vesicles: a novel platform for bacterial vaccines. Sci Rep 6(1):24931. https://doi.org/10.1038/srep24931
8. Rajagopal M, Walker S (2017) Envelope structures of gram-positive bacteria. Curr Top Microbiol Immunol 404:1–44. https://doi.org/10.1007/82_2015_5021
9. Schneewind O, Missiakas D (2016) Lipoteichoic acid synthesis and function in gram-positive bacteria. In: Geiger O (ed) Biogenesis of fatty acids, lipids and membranes. Springer International Publishing, Cham, pp 1–18. https://doi.org/10.1007/978-3-319-43676-0_17-1
10. Percy MG, Gründling A (2014) Lipoteichoic acid synthesis and function in gram-positive bacteria. Ann Rev Microbiol 68:81–100. https://doi.org/10.1146/annurev-micro-091213-112949
11. Wenzel CQ, Mills DC, Dobruchowska JM et al (2020) An atypical lipoteichoic acid from *Clostridium perfringens* elicits a broadly cross-reactive and protective immune response. J Biol Chem 295(28):17. https://doi.org/10.1074/jbc.RA119.009978
12. Behr T, Fischer W, Peter-Katalinić J, Egge H (1992) The structure of pneumococcal lipoteichoic acid. Improved preparation, chemical and mass spectrometric studies. Eur J Biochem 207(3):1063–1075. https://doi.org/10.1111/j.1432-1033.1992.tb17143.x
13. Nariya H, Miyata S, Suzuki M, Tamai E, Okabe A (2011) Development and application of a method for counterselectable in-frame deletion in *Clostridium perfringens*. Appl Environ Microbiol 77(4):1375–1382. https://doi.org/10.1128/aem.01572-10
14. Sanchez CJ, Kumar N, Lizcano A et al (2011) *Streptococcus pneumoniae* in biofilms are unable to cause invasive disease due to altered virulence determinant production. PLoS One 6(12):e28738. https://doi.org/10.1371/journal.pone.0028738
15. Laemmli UK (1970) Cleavage of structural proteins during the assembly of the head of bacteriophage T4. Nature 227(5259):680–685. https://doi.org/10.1038/227680a0
16. Tsai CM, Frasch CE (1982) A sensitive silver stain for detecting lipopolysaccharides in polyacrylamide gels. Anal Biochem 119(1):115–119. https://doi.org/10.1016/0003-2697(82)90673-x
17. Varki A, Cummings RD, Aebi M, Packer NH, Seeberger PH, Esko JD, Stanley P, Hart G, Darvill A, Kinoshita T, Prestegard JJ, Schnaar RL et al (2015) Symbol nomenclature for graphical representations of glycans. Glycobiology 25(12):1323–1324. https://doi.org/10.1093/glycob/cwv091

Chapter 11

Microbial Lectin-Targeting Mucus: A Glycobioinformatic Approach

Frédérique Lisacek, Boris Schnider, David Ohayon, and Anne Imberty

Abstract

The motifs exposed on surface glycans of mucins serve as receptors to a range of proteins expressed by microorganisms. The specificity of microbial glycan-binding proteins, i.e., lectins, towards human mucin epitopes is the result of co-evolution. The binding of microbes to mucins is described in several databases hosted on the UniLectin portal. In particular, UniLectin3D provides structural details on the recognition process and LectomeXplore allows for the identification of putative lectins in the genomes of mucin-associated microbiome species. The usage of these resources is illustrated in this chapter.

Key words Adhesin, Carbohydrate recognition domain, Database, 3D structure, Ligand

1 Introduction

Mucins create a thick glycan barrier on the surface of epithelial cells and are the primary components of mucus that coats epithelial tissues. This mucus barrier effectively prevents most potential infections through mucus clearance and is a major contributor to immunity. However, in the context of host-pathogen interactions, bacteria and viruses have developed strategies during evolution to enhance virulence, develop resistance, and escape immune systems, while hosts have strengthened immune responses and acquired tolerance mechanisms [1, 2].

In this fight spread over millions of years, protein-carbohydrate interactions play a dual and crucial role, with bacterial/fungal cell wall polysaccharides interacting with host receptors (lectins) and host O-glycans densely presented on mucins interacting with pathogen receptors. Consequently, pathogens evolved a large variety of specialized glycan binding proteins, in the form of adhesins that can be part of organelles such as pili or flagella, soluble lectins, and toxins [3]. Several such pathogen receptors significantly contribute to the infection process. For example, *Helicobacter pylori* can

Inka Brockhausen (ed.), *Dynamics of Bacteria-Mucus Interactions*, Methods in Molecular Biology, vol. 2942,
https://doi.org/10.1007/978-1-0716-4627-4_11,

trigger gastric disorder potentially leading to gastric carcinoma, through the release of the BabA and SabB adhesins that mediate the interaction with gastric mucins: BabA binds fucosylated blood group oligosaccharides and SabB sialylated ones [4]. Fucosylated histo-blood group antigens on human intestinal mucus also serve as ligands for the capsid protein of noroviruses, the most common cause of gastroenteritis [5].

With the recent extensive focus on microbiomes, the variety of organisms living and foraging human mucins is getting more precise. However, most studies target the gut microbiome [6], with less information available on other organs, despite the knowledge of pathogen receptors required for the design of novel compounds, such as glycomimetics, expected to address the antibiotic resistance crisis [7].

Several glycobioinformatics databases and tools are available to investigate protein-glycan interactions. A collection of pathogen lectins and their ligands is recorded and annotated in SugarBindDB [8]. Three-dimensional information characterizing protein-carbohydrate interactions can be searched in UniLectin3D [9], and proteome-wide prediction of bacterial carbohydrate-binding can be explored with LectomeXplore [10]. We previously demonstrated how the prediction of putative lectins can be used for understanding commensal and pathogen colonization of the vaginal microbiome [11].

In the present chapter, a first example illustrates the search for known structures of lectins interacting with mucin oligosaccharides and the analysis of binding properties, while two other ones demonstrate how to search for new putative lectins in available genomes.

2 Materials

The UniLectin portal (unilectin.unige.ch) hosts several databases in which a wide range of information covering knowledge on lectins is accessible. The portal is freely available on the web, with visualization adapted for computer screen, tablets, and smartphones. In this chapter, we rely on two modules of UniLectin, namely, UniLectin3D and LectomeXplore.

The UniLectin3D database contains structural data on lectins and their interactions with carbohydrate ligands. The database primarily focuses on three-dimensional data sourced from the Protein Data Bank (PDB) [12], with careful curation of glycan topology. It proposes a curated classification based on fold and sequences, resulting in 110 classes. For each entry, literature and functional data such as specificity, are available, as well as links to specialized glyco-related databases. Additionally, it allows for 3D visualization of contacts between lectins and ligands through the

Protein-Ligand Interaction Profiler (PLIP) application [13]. As of April 2024 (release 1.5.1), the database includes 2635 manually curated 3D structures of lectins, 1619 of them in complex with a ligand, and spans 686 different lectins. Several lectin 3D structure search options are proposed on the UniLectin3D home page. Browsing is accessible through three sunbursts based on lectin fold classification, biological origin (taxonomy), and monosaccharide composition of glycan ligand. A hierarchical taxonomic tree can also be used to explore the classification. The LectomeXplore database includes putative lectins (lectomes) identified by screening translated genomes (proteomes) from UniProtKB and NCBI-nr, using Hidden Markov Model (HMM)-based profiles of the 110 classes of UniLectin3D [10]. All lectins (lectome) can be searched in a single organism and possibly compared across different organisms (phila, environment, pathogenic state, etc.). LectomeXplore also highlights the occurrence of a lectin class in all scanned translated genomes. When available, predicted lectin domains are also associated with protein families as identified in the Pfam [14] and InterPro [15] databases.

3 Methods

3.1 Searching for 3D Structures of Bacterial Lectins in Complex with a Defined Mucin Oligosaccharide

In order to target mucin-specific lectins, we start from the mucin glycan composition, and to that end, click on the "Glycan search" button.

1. *Selecting Mucin Oligosaccharide Fragments in Lectin 3D Structures*

   The "glycan search" button on the UniLectin3D home page prompts the graphical representation of the 26 monosaccharides present in the lectin/glycan complexes of the database, using SNFG representation [16] (Fig. 1). Since GalNAc is the monosaccharide attached on Serine and threonine in O-glycans, the D-GalpNAc box is selected to start the analysis of complexes.

   Clicking on D-GalpNAc triggers the display of the 41 different GalNAc-containing ligands of the database. These are not limited to O-glycans, since GalNAc is also a constituent of oligosaccharide moieties of glycosphingolipids and of A-type blood group antigens present on all types of glycans. Nevertheless, as shown in Fig. 1, some GalNAc-containing motifs specific to O-glycans are identified such as Core 1 (T-antigen Galβ1-3GalNAc) and sialylated forms (NeuAcα2-3Galβ1-3GalNAc and NeuAcα2-3Galβ1-3(NeuAcα2-6)GalNAc) and Core 2 (3Galβ1-3(GlcNAcβ1-6)GalNAc), SLex-Core 2 (NeuAc(α2-3)Gal(β1-4)[Fuc(α1-3)]GlcNAc(β1-6)[Gal(β1-3)]GalNAc) and sialyl Tn-antigen (NeuAc(α2-6)GalNAc).

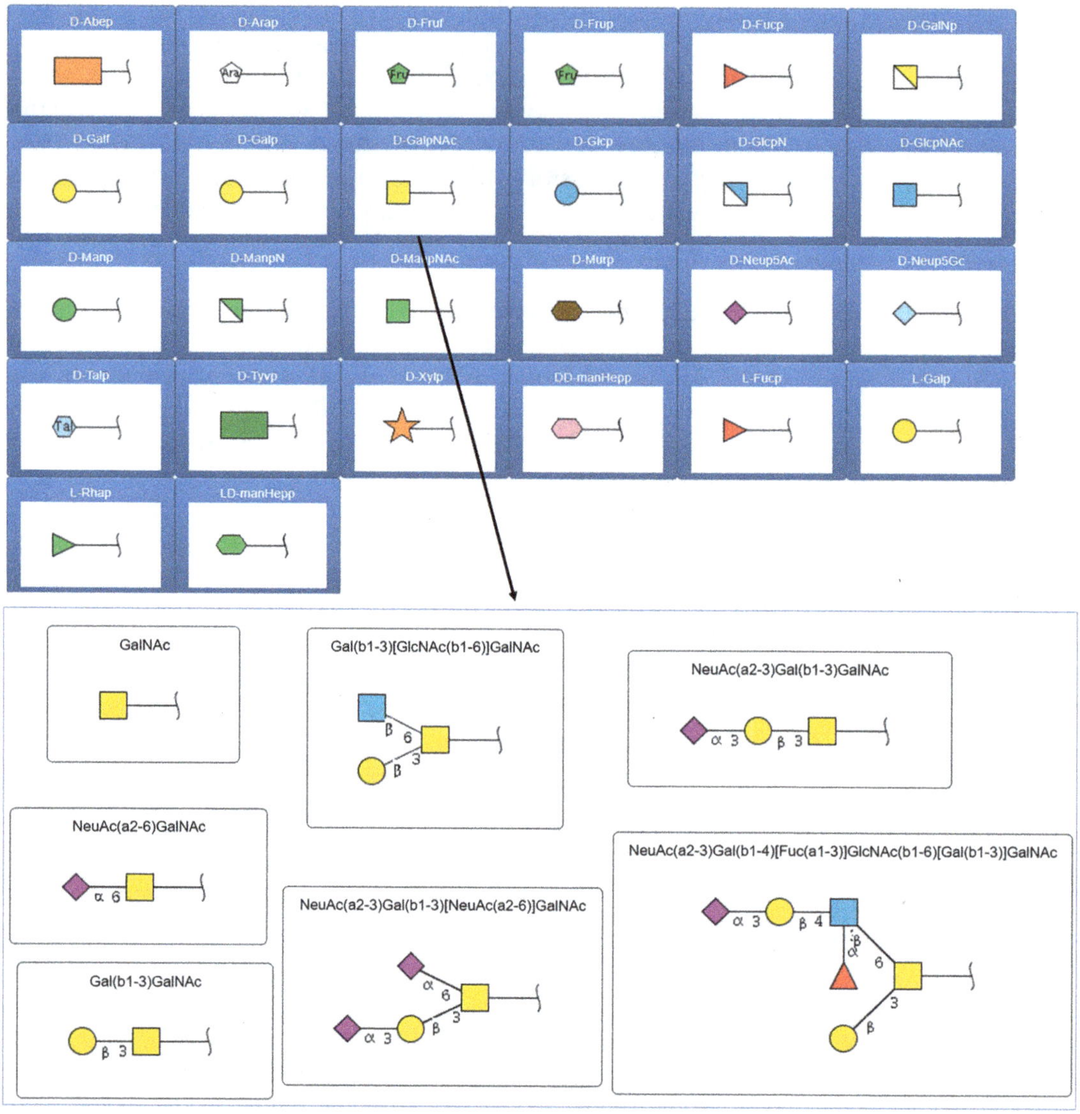

Fig. 1 Top: collection of monosaccharides available in the glycan search option of UniLectin3D. Clicking on the D-GalNAcp button turns its background green and triggers the onset of 41 glycan structures. Bottom: selection of seven oligosaccharides that contain GalNAc and are constituents of O-linked glycans

Selecting one of the motifs, for example, sialyl T-antigen (NeuAcα2–3Galβ1–3GalNAc), automatically opens the "search by field" window with a pre-filled "IUPAC condensed" field.

2. *Refining Search for Bacterial Lectins Binding to an Exact Motif of Mucins*

Click on the "Explore X-ray structures" button in the "Search by field" window of UniLectin3D, filled with NeuAc(a2–3)Gal (b1–3)GalNAc in the "IUPAC condensed" search field. This

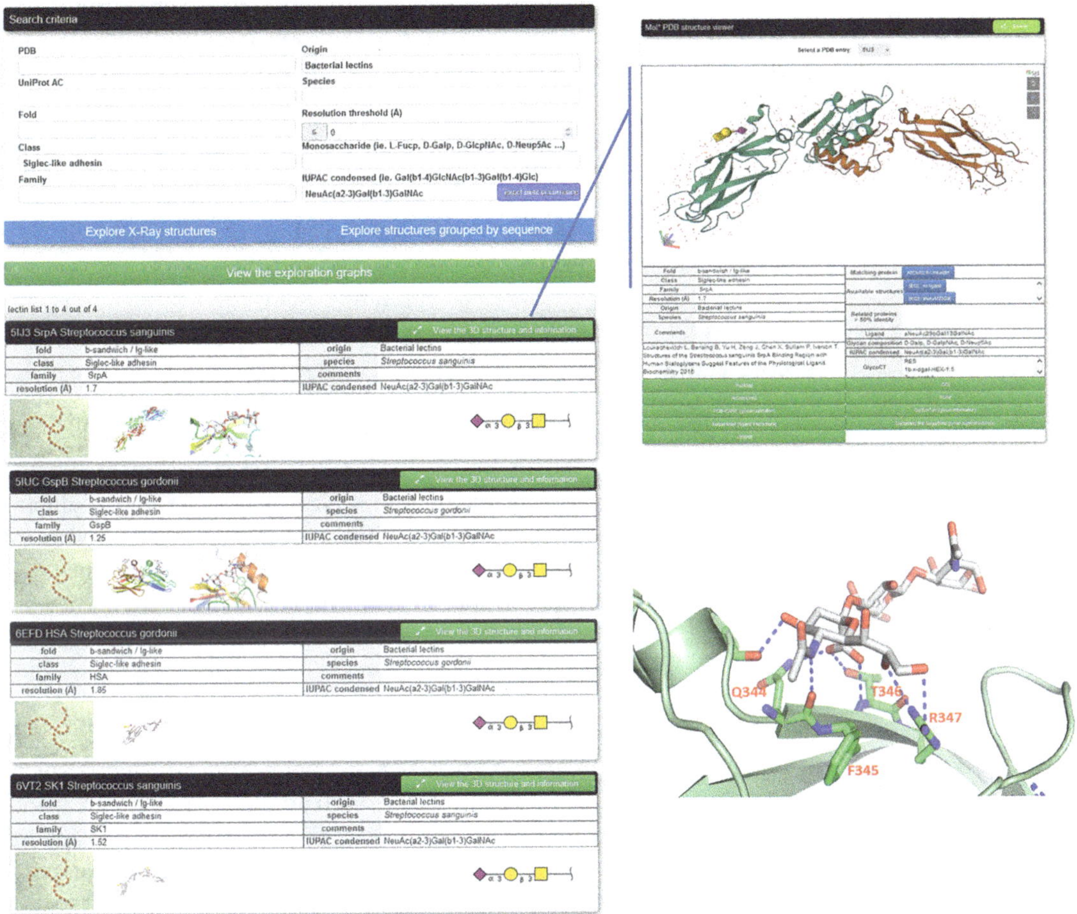

Fig. 2 Left: results of searching for the exact motif NeuAcα2-3Galβ1-3GalNAc combined with bacterial origin in UniLectin3D. Right top: result of selecting the 5IJ3 structure and opening the corresponding entry. Right bottom: graphical representation of the mucin oligosaccharide binding site with the Pymol software after downloading the coordinates of the PDB file

results in showing the 21 lectin entries of viral and bacterial origin matching this query. The search can be refined using additional criteria, such as specifying "bacterial lectins" in the "Origin" field. The displayed entries contain different type of glycans encompassing O-glycans as well as glycolipids from ganglioside families. To ensure the selection is restricted to complexes with the searched trisaccharide, press the "exact motif occurrence" button next to the "IUPAC condensed" search field (Fig. 2).

The results are then limited to four entries of bacterial lectins complexed with sialyl T-antigen, all belonging to the Siglec-like adhesin class identified in four serine-rich repeat proteins (SRRPs), SrpA and SK1 from *Streptococcus sanguinis* and GspB and Hsa from *Streptococcus gordonii* (Fig. 2).

SRRPs (also called serine-rich repeat adhesins) are large glycoprotein adhesins expressed in many Gram-positive bacteria [17] and consist of a variety of binding domains [18]. Some streptococci SRRs contain a Siglec-like adhesin domain with the same fold and specificity as the animal I-type lectins [19]. In UniLectin3D, the four entries are listed with images, classification details, and a representation of the glycan ligand. Additional information can be obtained clicking on the "View the 3D structure and information" button (Fig. 2). This opens a new window with detailed X-ray data interactively displayed as a 3D structure with the Mol* viewer software [20]. The interface also contains a number of buttons connecting to popular external databases such as PubMed (https://pubmed.ncbi.nlm.nih.gov/) or UniProt [21], as well as more specific glycobioinformatics resources related to the protein structure and specificity and ligand (right panel in Fig. 2). Finally, detailed information about the interaction between the protein and the glycan ligands is available for inspection at atomic level in the "Swiss Model PLIP interaction viewer" window, where the PDB file can be downloaded and further analyzed with any other 3D structure visualization software (e.g., PyMol (Schrödinger) in (Fig. 2). In the case of SrpA (PDB code 5IJ3), the specific recognition of the terminal NeuAc is clearly due to the QFTR peptide (position 344–347). In this stretch of four amino acids, hydrophobic contact is generated with F345 aromatic ring inserted between the glycerol and methyl groups of NeuAc, while hydrogen bonds [1] fork between R347 side chain and glycerol hydroxyl groups and [2] bind side chains of Q344 and T346 hydrogen to the acidic group of NeuAc.

3.2 Mining Lectins from Mucin-Associating Bacteria with LectomeXplore: Evaluating the Occurrence of Siglec-like Adhesins in Bacterial Kingdom

The LectomeXplore homepage offers several search options. Using "Search by field" "Siglec-like adhesin" can be selected in the "lectin class" field. The default score threshold of 0.5 (*see* **Note 1**) is rather stringent and will limit the results to 181 proteins from a few Streptococci species, mostly *S. gordonii* and *S. sanguinis*. Lowering the score threshold to 0.4 is preferable for exploring lectomes. In the present case, results are expanded to 289 putative SRR adhesins (Fig. 3). While most of them belong to Streptococci (198), a large number (63) are identified in *Gardnerella* and almost exclusively (59) in *Gardnerella vaginalis*. Interestingly, this pathogen responsible for bacterial vaginosis [22] has been demonstrated to deplete mucosal sialic acid, reducing intact sialoglycans in vaginal secretion [23].

Coming back to the "search by field" page and inputting "*Gardnerella vaginalis*" in the "Species" field launch a new search that targets the 59 entries mentioned above. Each can be viewed with details of the whole protein sequence and the putative lectin domain. Selecting, for example, UniProt A0A133NU91 (score

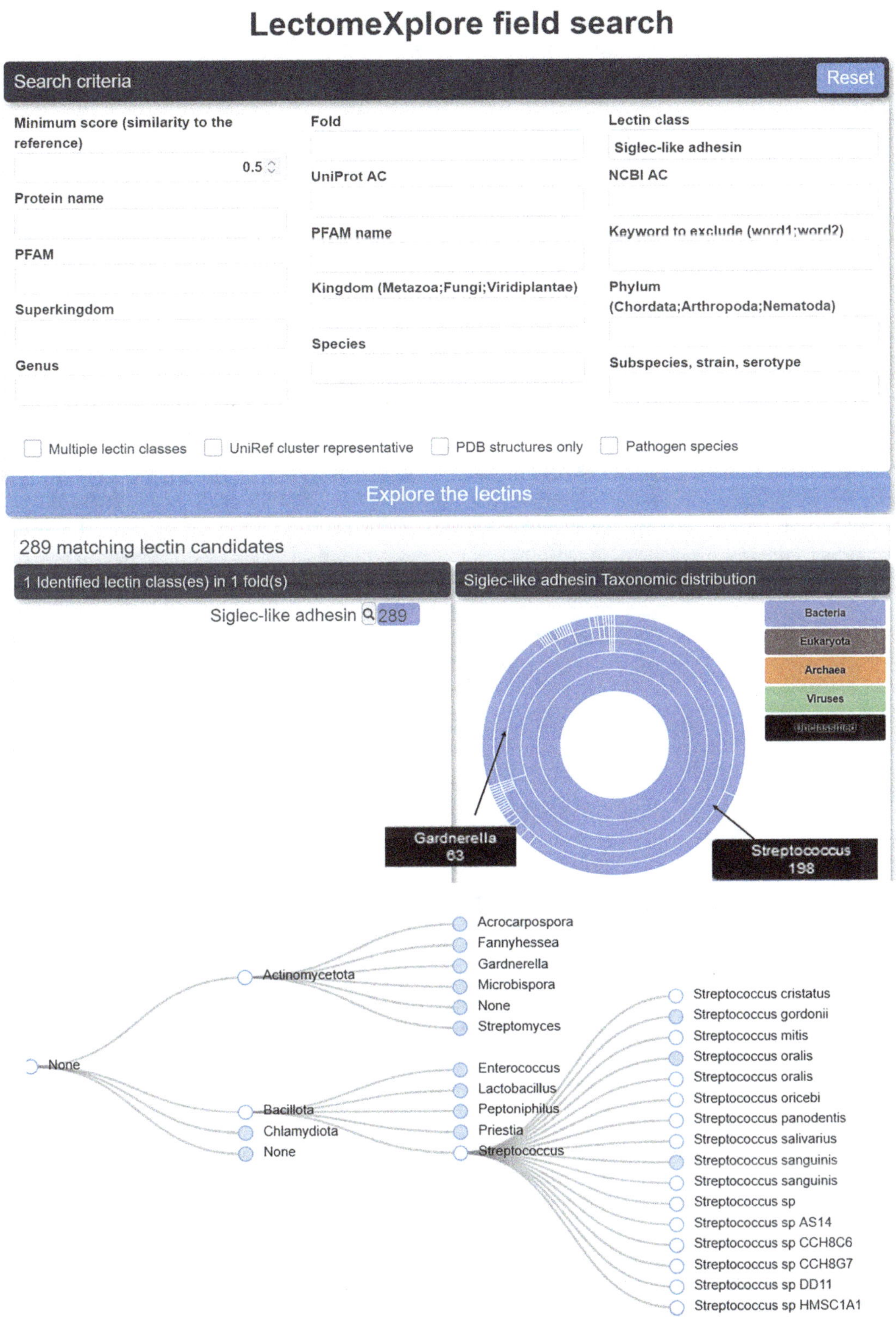

Fig. 3 Top: results of searching for Siglec-like adhesins with a minimum score of 0.4 in LectomeXplore. Right top: result of selecting the 5IJ3 structure and opening the entry. Right bottom: graphical representation

0.456) and clicking on the "View information" button open a new window, which shows a fragment of a very large protein (2841amino acids), annotated as "Rib/alpha/Esp surface antigen repeat-containing domain protein" consisting of many Rib domains (Pfam PF08428 (*see* **Note 2**)), and three Siglec-like domains (GspA/SrpA SigLec-like domain in Pfam PF20164) (Fig. 4). The local alignment of one of the latter domains with the corresponding Siglec-like adhesin reference from UniLectin3D confirms sequence similarity, and the conservation of the stretch of amino acids binding sialic acid (EYTRTI in A0A133NU91 compared to QFTRTI in the reference) thereby maintaining the hydrophobic contact and hydrogen bond possibilities. It would be of interest to check the ability of this protein to bind sialic acid since the presence of a negatively charged amino acid (Glu instead of Gln) may interfere.

3.3 Exploring with LectomeXplore the Lectome of a Bacterium from the Oral Cavity Interacting with Salivary Mucins

Streptococcus mitis is a normal resident of the oral cavity that uses salivary mucins for adhesion and food source [24]. Searching for this species in UniLectin3D results in the identification of a Siglec-like adhesin part of a Hsa-like SRR [19] and a F-type lectin domain, part of lectinolysin, a cholesterol-dependent toxin that causes cytolysis by forming pores in host membranes [25]. Several crystal structures of this fucose binding domain are described in complexes with a variety of histo-blood group oligosaccharides present on mucins [25].

In addition to these well-characterized glycan binding proteins, other members of the *S. mitis* lectome can be predicted by searching in LectomeXplore. The results obtained with a score threshold of 0.4 are displayed in Fig. 5. Two other classes of putative lectins are identified in addition of the ones above. Toxin-repetitive domains, present in *Clostridium difficile* TcdA, bind to glycans on intestinal epithelial cells with specificity for αGal containing oligosaccharides [26]. In Streptococci, these repeats are found in association with several enzyme domains and described as cell-wall binding domain or choline-binding domain [27]. A role in mucin-binding is therefore possible but is awaiting experimental confirmation. Factor H-binding protein, also called SapD, is a *Streptococcus suis* adhesin responsible of emerging zoonosis that causes meningitis in pigs and more rarely in humans [28]. Interestingly, the specificity is also directed to αGal containing oligosaccharides. This motif can be present in mucins as part as blood group epitopes, especially for nonhuman mammals, as Galili antigen. Our search therefore resulted in proposing two putative lectins in *Streptococcus mitis*, i.e., TcdA- and SapD-related glycan-binding

Fig. 3 (continued) of the mucin. Bottom: resulting hierarchical taxonomy tree with selected expansions of the two most populated species, i.e., *G. vaginalis* and *S. sanguinis*

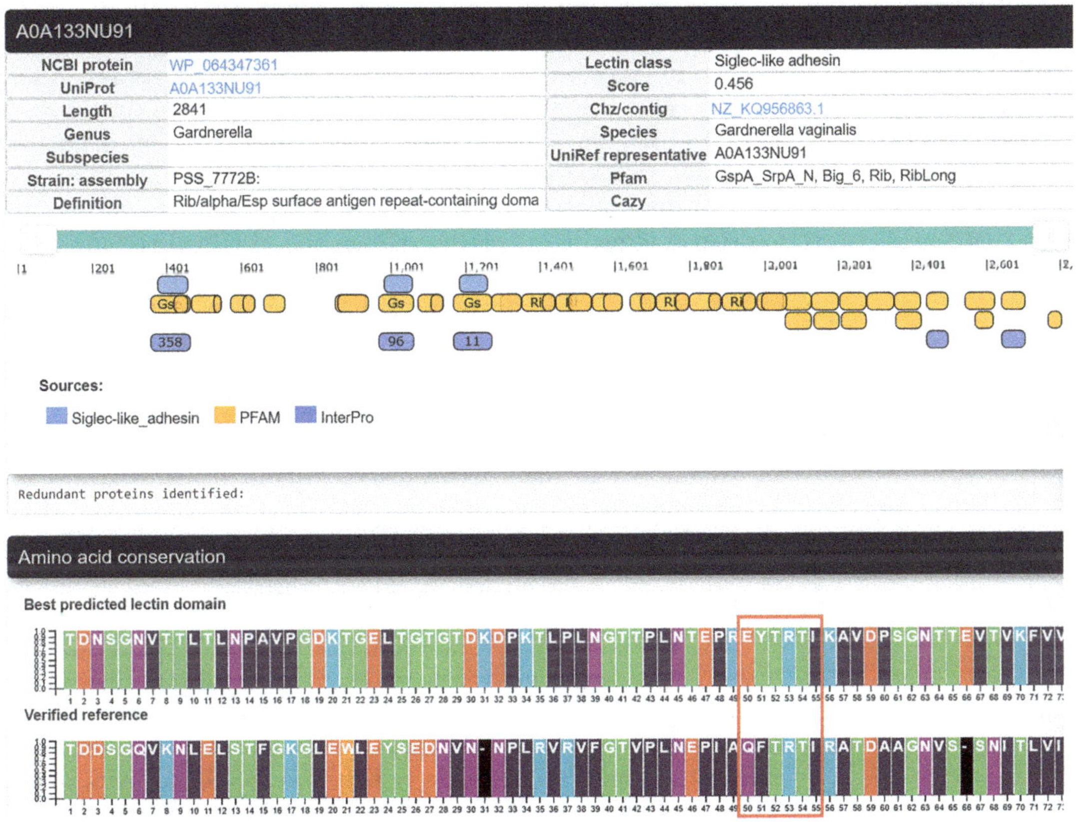

Fig. 4 Partial view of the LectomeXplore entry of one of the Siglec-like adhesins predicted to occur in the lectome of *G. vaginalis*. The conserved amino acids that shape the sialic acid binding site are boxed in red

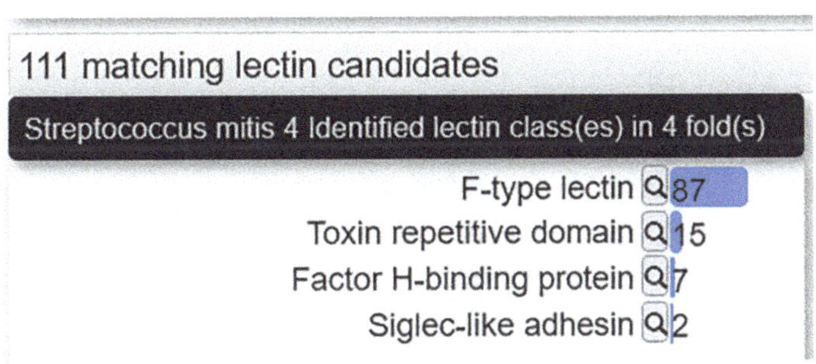

Fig. 5 Results of searching *Streptococcus mitis* with a score threshold of 0.4 in LectomeXplore

domains, in addition to the two that were identified previously. Again, experimental work would be needed to confirm the binding properties of these proteins.

3.4 Comparing LectomeXplore Predicted Lectomes in Several Streptococci Species of the Oral Cavity

The expected Streptococci residents of the oral cavity have been categorized in five groups: [1] mutans group (e.g., *Streptococcus mutans* and *sobrinus*), [2] salivarius group (e.g., *Streptococcus salivarius*), [3] anginosus group (*Streptococcus anginosus* and *intermedius*), [4] sanguinis group (*Streptococcus sanguinis* and *gordonii*), and [5] mitis group (*Streptococcus mitis* and *oralis*) [29]. The genomes of these species have been sequenced, and consequently, the translated sequences have been screened by the LectomeXplore engine and putative lectins are stored in the LectomeXplore database. Up to ten lectomes can be compared. The following ten species are considered:

1. Mutans group: *Streptococcus mutans* and *sobrinus.*
2. Salivarius group: *Streptococcus salivarius.*
3. Anginosus group: *Streptococcus anginosu*, *constellatus*, and *intermedius.*
4. Sanguinis group: *Streptococcus sanguinis* and *gordonii.*
5. Mitis group: *Streptococcus mitis* and *oralis.*

In the homepage of LectomeXplore, find and click on the "Lectin class heatmap" button. Lower the threshold value to 0.2 (*see* **Note 3**), and type one at a time the ten streptococcal species' name (autocompletion helps get to it faster) clicking the "Add" button each time. When the ten names are recorded, click on the "Compare" button. After a short while, the heatmap featuring each species' name and each fold name of the first level of classification (*see* **Note 4**) is displayed. Figure 6 shows this result and highlights a specific profile for each group.

4 Notes

1. The lectin prediction score in LectomeXplore is calculated and normalized on the basis of the e-value generated by the hidden Markov Model and sequence identity limited to the lectin domain (i.e., not the whole sequence). In practice, a 0.5 threshold is conservative, and lowering to 0.4 does not significantly increase the number of false positives. This number is certainly higher when selecting a number lower than 0.4, yet there is no ground for setting an ideal minimum.
2. The Pfam database is no longer accessible through the original website, but all information was transferred to the InterPro database to preserve the validity of Pfam accession numbers.
3. The 0.2 value is low and could retrieve dubious lectins. Nonetheless, for a first estimate, the distinction between species may be easier to observe despite possible noise.

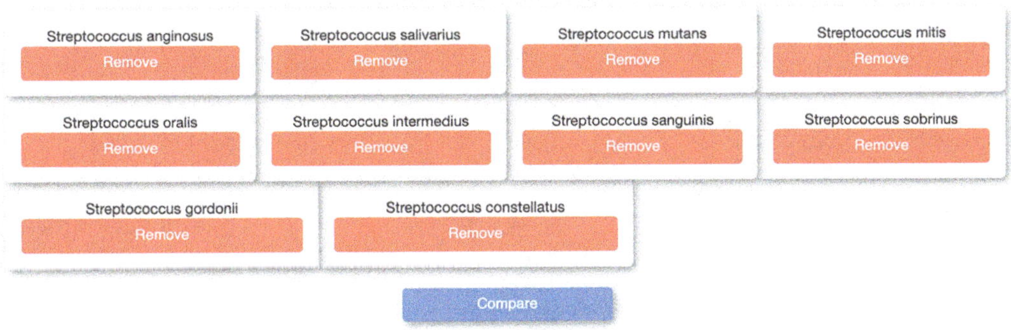

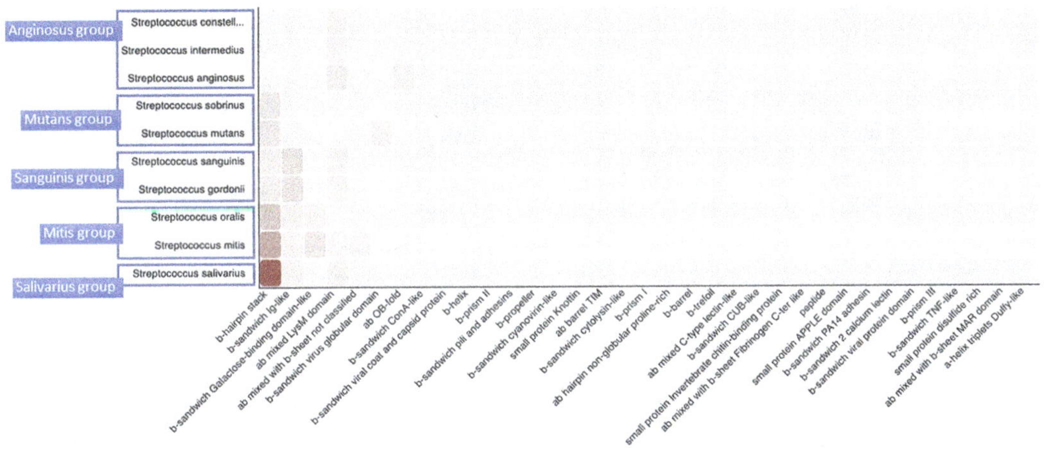

Fig. 6 Heatmap showing the lectome profiles of 10 Streptococcus species known to reside in the oral cavity. These species are shown with their group membership. The heatmap highlights how each bacterial species is equipped with a lectin arsenal that is typical of its group

4. In the default settings, lectome comparison is based on content classified by "Fold." To obtain a more specific comparison, change "Fold" to "Class" at the top of the page.

References

1. Bergstrom K, Xia L (2022) The barrier and beyond: roles of intestinal mucus and mucin-type O-glycosylation in resistance and tolerance defense strategies guiding host-microbe symbiosis. Gut Microbes 14(1):2052699. https://doi.org/10.1080/19490976.2022.2052699
2. Morozov V, Borkowski J, Hanisch FG (2018) The double face of mucin-type O-glycans in lectin-mediated infection and immunity. Molecules 23(5). https://doi.org/10.3390/molecules23051151
3. Moonens K, Remaut H (2017) Evolution and structural dynamics of bacterial glycan binding adhesins. Curr Opin Struct Biol 44:48–58. https://doi.org/10.1016/j.sbi.2016.12.003
4. Matos R, Amorim I, Magalhaes A, Haesebrouck F, Gartner F, Reis CA (2021) Adhesion of helicobacter species to the human gastric mucosa: a deep look into glycans role. Front Mol Biosci 8:656439. https://doi.org/10.3389/fmolb.2021.656439
5. Taube S, Mallagaray A, Peters T (2018) Norovirus, glycans and attachment. Curr Opin

Virol 31:33–42. https://doi.org/10.1016/j.coviro.2018.04.007

6. Leviatan S, Shoer S, Rothschild D, Gorodetski M, Segal E (2022) An expanded reference map of the human gut microbiome reveals hundreds of previously unknown species. Nat Commun 13(1):3863. https://doi.org/10.1038/s41467-022-31502-1
7. Leusmann S, Menova P, Shanin E, Titz A, Rademacher C (2023) Glycomimetics for the inhibition and modulation of lectins. Chem Soc Rev 52(11):3663–3740. https://doi.org/10.1039/d2cs00954d
8. Mariethoz J, Khatib K, Alocci D, Campbell MP, Karlsson NG, Packer NH, Mullen EH, Lisacek F (2016) SugarBindDB, a resource of glycan-mediated host-pathogen interactions. Nucleic Acids Res 44(D1):D1243–D1250. https://doi.org/10.1093/nar/gkv1247
9. Bonnardel F, Mariethoz J, Salentin S, Robin X, Schroeder M, Perez S, Lisacek F, Imberty A (2019) UniLectin3D, a database of carbohydrate binding proteins with curated information on 3D structures and interacting ligands. Nucleic Acids Res 47:D1236–D1244. https://doi.org/10.1093/nar/gky832
10. Bonnardel F, Mariethoz J, Perez S, Imberty A, Lisacek F (2021) LectomeXplore, an update of UniLectin for the discovery of carbohydrate-binding proteins based on a new lectin classification. Nucleic Acids Res 49(D1):D1548–D1554. https://doi.org/10.1093/nar/gkaa1019
11. Bonnardel F, Haslam SM, Dell A, Feizi T, Liu Y, Tajadura-Ortega V, Akune Y, Sykes L, Bennett PR, MacIntyre DA, Lisacek F, Imberty A (2021) Proteome-wide prediction of bacterial carbohydrate-binding proteins as a tool for understanding commensal and pathogen colonisation of the vaginal microbiome. Biofilms Microbiomes 7:49. https://doi.org/10.1038/s41522-021-00220-9
12. Burley SK, Bhikadiya C, Bi C, Bittrich S et al (2021) RCSB protein data bank: powerful new tools for exploring 3D structures of biological macromolecules for basic and applied research and education in fundamental biology, biomedicine, biotechnology, bioengineering and energy sciences. Nucleic Acids Res 49(D1):D437–D451. https://doi.org/10.1093/nar/gkaa1038
13. Salentin S, Schreiber S, Haupt VJ, Adasme MF, Schroeder M (2015) PLIP: fully automated protein-ligand interaction profiler. Nucleic Acids Res 43(W1):W443–W447. https://doi.org/10.1093/nar/gkv315
14. Mistry J, Chuguransky S, Williams L et al (2021) Pfam: the protein families database in 2021. Nucleic Acids Res 49(D1):D412–D419. https://doi.org/10.1093/nar/gkaa913
15. Paysan-Lafosse T, Blum M, Chuguransky S et al (2023) InterPro in 2022. Nucleic Acids Res 51(D1):D418–D427. https://doi.org/10.1093/nar/gkac993
16. Neelamegham S, Aoki-Kinoshita K, Bolton E et al (2019) Updates to the symbol nomenclature for glycans guidelines. Glycobiology 29(9):620–624. https://doi.org/10.1093/glycob/cwz045
17. Cinar MS, Niyas A, Avci FY (2024) Serine-rich repeat proteins: well-known yet little-understood bacterial adhesins. J Bacteriol 206(1):e0024123. https://doi.org/10.1128/jb.00241-23
18. Latousakis D, MacKenzie DA, Telatin A, Juge N (2020) Serine-rich repeat proteins from gut microbes. Gut Microbes 11(1):102–117. https://doi.org/10.1080/19490976.2019.1602428
19. Bensing BA, Stubbs HE, Agarwal R et al (2022) Origins of glycan selectivity in streptococcal Siglec-like adhesins suggest mechanisms of receptor adaptation. Nat Commun 13(1):2753. https://doi.org/10.1038/s41467-022-30509-y
20. Sehnal D, Bittrich S, Deshpande M et al (2021) Mol* viewer: modern web app for 3D visualization and analysis of large biomolecular structures. Nucleic Acids Res 49(W1):W431–W437. https://doi.org/10.1093/nar/gkab314
21. UniProt Consortium (2021) UniProt: the universal protein knowledgebase in 2021. Nucleic Acids Res 49(D1):D480–D489. https://doi.org/10.1093/nar/gkaa1100
22. Shvartsman E, Hill JE, Sandstrom P, MacDonald KS (2023) Gardnerella revisited: species heterogeneity, virulence factors, mucosal immune responses, and contributions to bacterial vaginosis. Infect Immun 91(5):e0039022. https://doi.org/10.1128/iai.00390-22
23. Agarwal K, Lewis AL (2021) Vaginal sialoglycan foraging by gardnerella vaginalis: mucus barriers as a meal for unwelcome guests? Glycobiology 31(6):667–680. https://doi.org/10.1093/glycob/cwab024
24. Okahashi N, Nakata M, Kuwata H, Kawabata S (2022) Oral mitis group streptococci: a silent majority in our oral cavity. Microbiol Immunol 66(12):539–551. https://doi.org/10.1111/1348-0421.13028
25. Feil SC, Lawrence S, Mulhern TD et al (2012) Structure of the lectin regulatory domain of the cholesterol-dependent cytolysin lectinolysin reveals the basis for its Lewis antigen specificity.

Structure 20(2):248–258. https://doi.org/10.1016/j.str.2011.11.017

26. Greco A, Ho JG, Lin SJ, Palcic MM, Rupnik M, Ng KK (2006) Carbohydrate recognition by Clostridium difficile toxin A. Nat Struct Mol Biol 13(5):460–461. https://doi.org/10.1038/nsmb1084
27. Swiatlo E, Champlin FR, Holman SC, Wilson WW, Watt JM (2002) Contribution of choline-binding proteins to cell surface properties of Streptococcus pneumoniae. Infect Immun 70(1):412–415. https://doi.org/10.1128/IAI.70.1.412-415.2002
28. Zhang C, Hao H, Yu Y et al (2016) Structural basis of the interaction between the meningitis pathogen Streptococcus suis adhesin fhb and its human receptor. FEBS Lett 590(9):1384–1392. https://doi.org/10.1002/1873-3468.12174
29. Kreth J, Merritt J, Qi F (2009) Bacterial and host interactions of oral streptococci. DNA Cell Biol 28(8):397–403. https://doi.org/10.1089/dna.2009.0868

Chapter 12

Exploring the *O*-glycomic Degradome Using Natural Mucin Libraries

Marthe Tofthagen, Chunsheng Jin, Piyush Patel, Fei Sjöberg, Ana S. Luis, Marissa L. Maciej-Hulme, and Niclas G. Karlsson

Abstract

The interactions between intestinal mucins and the gut microbiota significantly influence digestive processes, metabolic activities, and pathogen resistance in the gastrointestinal tract. To explore this dynamic relationship, this chapter describes a workflow for studying the degradation of mucin oligosaccharides from isolated mucins. The approach involves purifying mucins from intestinal tissues and then subjecting them to a mixture of fecal glycosidases originating from commensal bacteria. The remaining oligosaccharides attached to the mucins after glycosidase treatment are released and then characterized and quantified using liquid chromatography-mass spectrometry (LC-MS)^2. This approach identifies relevant intestinal oligosaccharides that are degraded by commensal flora. By combining natural oligosaccharide characterization with measurement of their degradation, we expand the repertoire of glycan structures for studying the intestinal glycomic degradome. Few commercially available characterized glycans or glycoproteins exist that cover the unique type of glycan mixtures found in the intestines where most microbes in the gastrointestinal tract reside. The presented workflow uses relevant glyco-substrates to investigate mucus degradation patterns, which are crucial for discerning pathological conditions associated with intestinal dysbiosis.

Key words Mucins, *O*-glycosylation, Microbiota, LC-MS, Chromatography, Mass spectrometry, Dot blotting, Mucin degradation, Exoglycosidase assay

1 Introduction

The intestines are important digestive organs. The mucosal barrier in the intestines protects the host from the intestinal bacteria. At the same time, the commensal bacteria and their interplay with the gut play a key role in digestion, absorption, and metabolism [1]. The human gastrointestinal tract is a highly colonized environment. This diverse community, known as the microbiota, has coevolved with the host to adapt and gain resistance to pathogens [2]. Mucins are the major proteins of the mucosa and are heavily glycosylated, where up to 80% of the molecular mass are made up of glycans (*O*-linked glycans) covalently linked to Ser/Thr residues

Inka Brockhausen (ed.), *Dynamics of Bacteria-Mucus Interactions*, Methods in Molecular Biology, vol. 2942,
https://doi.org/10.1007/978-1-0716-4627-4_12,

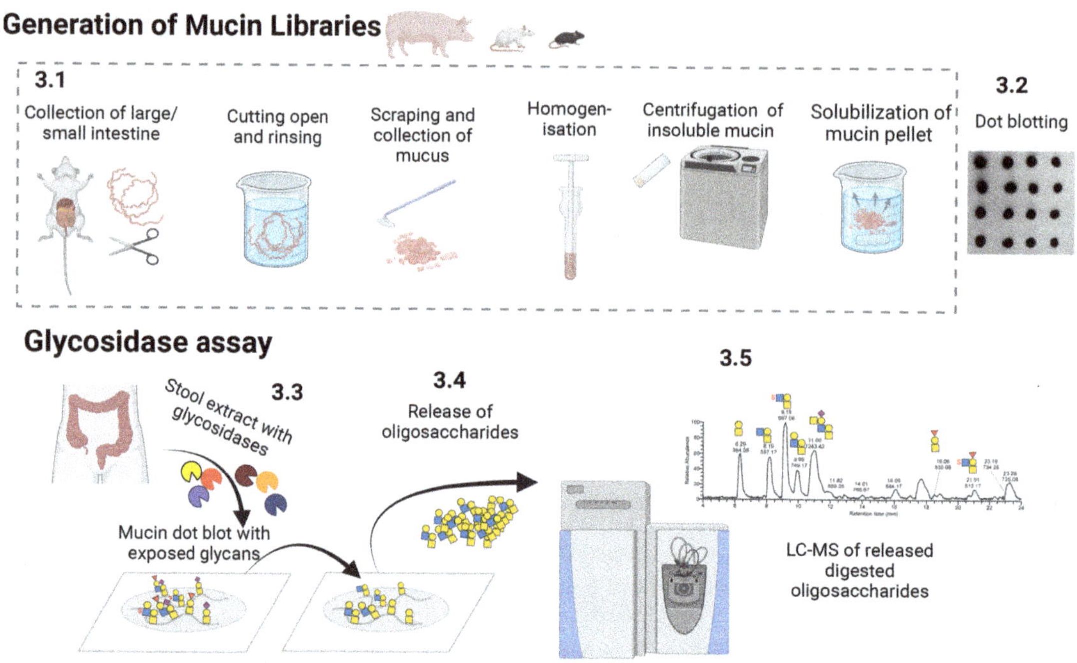

Fig. 1 Overview of the suggested workflow. The upper panel refers to the isolation of insoluble mucins and dot blotting of mucins for exoglycosidase digestion. This workflow for generation of mucin libraries includes collection of intestines, clean up, separation and solubilization (*3.1*), and generation of dot blots (*3.2*). Lower panel describes the workflow for measuring exoglycosidases, activity using dot blotted mucins (*3.3*), release of remaining oligosaccharides (*3.4*), and LC-MS analysis (*3.5*) Numbers in the figure refers to the different protocols in this chapter. (Figure generated in BioRender (www.biorender.com))

located mainly into what is known as mucin domains, where in addition to abundance Ser and Thr residues, the domains also are characterized by their high Pro content. *O*-glycans on intestinal mucins are involved in maintaining the balance of intestinal microflora, especially in the colon [3, 4]. Intestinal bacteria degrade mucins and their O-linked glycans. Degraded mucin glycans are then used in bacterial metabolism to support the growth and survival of commensal bacteria [5]. At the same time, mucins are providing a physical barrier which prevents the pathogens from infiltrating the mucosal layer. To study the interaction between mucins and bacteria, we have adopted an approach involving isolation of insoluble mucins from the intestines of animal sources and use them as mucin glycan libraries (Fig. 1, upper panel). These mucin glycan libraries can be used as an in vitro model to assess the efficiency of glycan degradation of fecal glycosidases from animal and human stool samples using LC-MS^{2}. (Fig. 1 lower panel).

The workflow relies on incubating mucins immobilized on PVDF membranes with fecal extracts, which contain commensal bacteria from the microbiome. Subsequent analysis of remaining oligosaccharides after incubation with fecal extracts provides insight into the specificity of the degradation. This analysis is carried out using LC-MS^2 after release of oligosaccharides from the membranes [6]. By incubating fecal extracts from participants from various populations, this methodology provides information about mucin oligosaccharide degradation patterns in comparative studies. Differences in degradation profiles between healthy and diseased states highlight mechanisms underlying mucin degradation-associated diseases [7, 8]. Negative ion LC-MS and LC-MS^2 with graphitized carbon is an effective method to separate and analyze *O*-linked oligosaccharides with high sensitivity [6].

2 Materials

2.1 Isolation of Insoluble Intestinal Mucins

1. Small or large intestine from mice, rats, or other mammalian of interest.
2. Scissors/knife for cutting the intestines.
3. Scraping tool (e.g., tissue slide of glass).
4. Scraping board.
5. 0.90% NaCl solution.
6. Extraction GuHCl buffer: 6.0 M guanidine hydrochloride, 5.0 mM EDTA, 0.010 M NaH_2PO_4, adjusted to pH 6.5 with NaOH.
7. *N*-Ethylmaleimide (NEM).
8. Phenylmethylsulfonyl fluoride (PMSF) in 2-propanol, 100 mM (made fresh).
9. Centrifuge (e.g., Beckman JA-20 rotor) with appropriate centrifuge tubes.
10. Beakers/tubes, marked accordingly.
11. Dithiothreitol (DTT).
12. Heater (water bath, heating block) for 37 °C.
13. Reduction GuHCl buffer: 6.0 M guanidine hydrochloride, 0.10 M Tris, 5 mM EDTA adjusted to pH 8.0 with HCl.
14. Dialysis tubing (> 30,000 MWCO, e.g., cellulose ester spectra/Por 50 kDa).
15. Lyophilizer.

2.2 Blotting of Mucin to Membranes

1. Dissolved insoluble mucin fraction from 3.1 (about 1.0 mg mucin dry weight/mL, weighed out using a balance).
2. Buffer for solubilization of mucin (e.g., phosphate buffer saline, PBS).
3. Dot-blot apparatus (e.g., Bio-Rad, Hercules, CA).
4. Vacuum pump/manifold.

5. Blotting paper.
6. Polyvinylidene fluoride (PVDF) membrane.
7. Milli-Q water.
8. Methanol.
9. Staining solution/Alcian Blue (Alcian Blue solution—25% ethanol, 10% acetic acid, 0.125% Alcian Blue).
10. Destaining solution (100% methanol).
11. Stirrer/shaker.
12. Tubes.
13. Knife/scissors.
14. Forceps.

2.3 Exoglycosidase Assaying of Mucin Glycans

1. Stool samples (fresh or frozen −20 °C for less than 1 week).
2. Exoglycosidase buffer: phosphate buffer saline PBS.
3. PBS-T (PBS with 0.1% Tween 20).
4. Mucin dot blots (from 3.2).
5. Incubator (37 °C).
6. Heater (water bath, heating block) for 95 °C.

2.4 Release and Clean-up of O-linked Oligosaccharides from Membranes

1. β-elimination solution: 50 mM NaOH, 0.50 M $NaBH_4$.
2. Methanol.
3. 1.0 M HCl.
4. Glacial acetic acid.
5. Desalting solution: 1% glacial acetic acid in methanol.
6. Oven set to 50 °C.
7. Vacuum centrifuge.
8. Cation exchange beads, AG50WX8 (BioRad).
9. Tabletop centrifuge.
10. Zip tips.

2.5 Analysis of Released Glycans After Glycosidase Treatment

1. LC-MS^2 instrument adopted for *O*-linked glycan analysis (Table 1).
2. Software for automatic and manual qualitative/quantitative interpretation of MS and MS^2 data (Table 2).

3 Methods

3.1 Isolation of Insoluble Intestinal Mucins

Prepare the "extraction GuHCl buffer" prior to starting this protocol. Approximately 5 mL is needed per gram of wet mucosa.

Table 1
Examples of LC-MS set up for compatible with analysis of exoglycosidase digested mucin *O*-glycans on blots

Chromatography	Dimensions	Flow rate	MS	Ion mode	Collision	References
Homemade graphitized carbon column with trap	3 µm particle diameter, 100 mm × 0.100 mm i.d.	600 nL/min	IT (Paul trap)	–	CID	[9]
Commercial graphitized carbon column	5 µm particle size, 180 µm internal diameter × 10 cm i.d.	4.0 µL/min	LTQ Orbitrap	–	HCD	[10]
Homemade graphitized carbon column	5 µm particle size, 180 um internal diameter × 10 cm i.d.	5–10 µL/min	IT (linear)	–	CID	[11]

Table 2
Useful references for methodology and evaluation of LC-MS^2 of *O*-linked oligosaccharides

References	Information
[12]	Statistical comparison between *O*-linked glycan samples analyzed by LC-MS
[11]	Statistical comparison between *O*-linked glycan samples analyzed by LC-MS. packing graphitized carbon column
[13, 14]	Description of the software GlycoWorkbench to annotate glycans from MS and MS/MS
[15]	Minimal reporting standards for an MS glycomic experiment (MIRAGE)
[16]	Preparing *O*-linked MS data for open access repositories (UniCarb-DR and GlycoPost)
[17]	GlyoMod, a tool for matching theoretical compositions of glycans with parent ions detected in MS
[18]	The introduction of GlycoPost repository to store raw MS data for glycomics
[19]	Softwares to generate quantitative and qualitative *O*-linked glycomic data using the experimental MS/MS fragment database UniCarb-DB
[20, 21]	Diagnostic ions for interpretation of *O*-linked glycan MS/MS
[22]	The curated database Glyconnect of identified structures and compositions reported in literature, with reference to GlycoMod and UniProt

1. Collect small intestines/colon from sacrificed animals.
2. Cut the intestines open with scissors and remove the majority of intestinal content mechanically using blunt tools without destroying the mucosa layer.
3. Wash the intestines in 0.9% NaCl solution twice, or until they look relatively clean.

4. Use the designated scraping tool to carefully scrape the inside lining of the small intestine/colon.
5. Add NEM (*N*-ethylmaleimide) to the "extraction GuHCl" (guanidium hydrochloride, final concentration 5.0 mM (62.55 mg/100 mL)) and dissolve.
6. Add PMSF (phenylmethylsulfonyl fluoride) from the 100 mM stock to the "extraction GuHCl" to make a final concentration of 5.0 mM (*see* **Note 1**).
7. Transfer the mucus scrapings to a Dounce homogenizer and add "extraction GuHCl buffer" with the added protease inhibitors (NEM and PMSF) (5 mL "extraction GuHCl buffer"/g wet mucosa).
8. Homogenize samples and transfer to tubes that can be closed/sealed or a beaker to leave in cold room (8 °C) on a rolling wheel or magnetic stirrer overnight.
9. Transfer the samples into centrifuge tubes of an appropriate volume.
10. Centrifuge at 40,000×*g* (18,000 rpm using JA-20 rotor) at 10 °C for 30 min.
11. Discard the fatty floating upper layer and the supernatant and retain the pellet.
12. Add equal volume of cold extraction GuHCl without NEM and PMSF the pellet beaker/tube as in **step 7**, and stir gently at 4 °C for at least 3 h.
13. Centrifuge as **step 10**. Discard the supernatant. Check that there is no other left-over residue on the pellet.
14. Repeat **steps 12** and **13** once, and discard the supernatant, repeat the steps until the supernatant is clear after centrifugation, it is normal to expect some debris and fibers in the pellet. It usually takes about five to six centrifugations before the supernatant is clear, depending on how clean the starting material is.
15. Estimate the volume of the remaining pellet.
16. Add "reduction GuHCl buffer" to the pellet (approximately 5 mL buffer/mL of estimated pellet volume).
17. Add DTT (dithiothreitol) to mucin preparation to make a final concentration of 10 mM DTT and reduce mucins at 37 °C for 2 h or for 45 min at 65 °C.
18. Block free thiol groups with the addition of NEM to a final concentration of 25 mM. Stir slowly 16–18 h in the dark at room temperature.
19. Centrifuge as in **step 10**. Collect the supernatant and discard the pellet (cell and tissue debris).

20. Rinse dialysis tubing with Milli-Q water.
21. Seal one end of the dialysis tubing.
22. Fill the dialysis tubing with the supernatant containing the solubilized mucins, and seal the other end.
23. Dialyze against excess of water (>20 times the dialysis volume each time), change water every 4–16 h for four times.
24. Empty the dialyzed samples from the dialysis tubing into vessels for lyophilization, freeze at −80 °C, and lyophilize until dry.

3.2 Blotting of Mucin to Membranes

1. Clean and dry dot blot apparatus.
2. Place the gasket support plate in the vacuum manifold.
3. Place the seal on top of the gasket support plate.
4. Cut a PVDF (polyvinylidene fluoride) membrane to an appropriate size. Use forceps when handling the membrane and immerse the membrane in 100% methanol for 2 min, followed by washing with Milli-Q water for 2 min. Soak a pre-cut filter paper in 100% methanol and soak in Milli-Q water for 2 min.
5. Place the wet filter paper on the gasket support plate.
6. Place the wet PVDF membrane on top of the filter paper.
7. Assemble the dot blot according to the manufacturer's instructions.
8. Pipette 20 μL of prepared mucin samples (1.0 mg/mL in PBS (phosphate buffered saline)) into each well of the dot blot (*see* **Note 2**).
9. Use the vacuum pump attached to the apparatus/PVDF membrane to pass the samples through the membrane.
10. Add 50 μL of water to each well with sample, wash the sides of the wells, and apply vacuum.
11. Disassemble the dot blot apparatus and remove the PVDF membrane.
12. Stain the PVDF membrane in an appropriate stain, (e.g., Alcian Blue) and destain accordingly.
13. Cut out the dot blots, add them into separate 2 mL tubes, and store at −20 °C until use.

3.3 Exoglycosidase Assaying of Mucin Glycans

1. Suspend the stool samples in PBS (phosphate buffered saline) (1.0 mL per 30 mg of wet stool sample).
2. Add approximately 150 μL of the solution (stool sample and PBS) to two mucin dot blots.
3. Incubate at 37 °C for 2–15 h (*see* **Note 3**).

4. For negative control, heat the solution (fecal sample and PBS) at 95 °C for 15 min before adding it to 2 mucin dot blots. For release and instrumental control, use two dot blots without fecal extraction incubation.
5. After incubation, discard the solution wash the dots with 1 mL PBS containing 0.1% Tween 20, repeat three times. (3 × 1 mL PBS w/0.1% Tween 20).
6. Wash the dot blots thoroughly with water (4 × 1.0 mL of water and agitation for 10 min each time).

3.4 Release of Glycans from Membranes

1. Place the two dot blots from each incubation into a microtiter well.
2. Add 20 μL of β-elimination solution to each tube (*see* **Note 4**).
3. Cover and incubate overnight (about 16 h) at 50 °C (*see* **Note 5**).
4. Neutralize solution by adding 1 μL glacial acetic acid to each well or tube. Solutions should immediately fizz due to production of H_2 gas.

Prepare cation exchange columns in advance by doing the following steps:

5. Add 25 μL of cation exchange suspension (50% Milli-Q, 50% methanol) to ZipTips (1 ZipTip per sample).
6. Place columns in 1.5 mL Eppendorf tubes and centrifuge for 10 s. Wash columns with 2 × 60 μL methanol and 2 × 60 μL 1.0 M HCl, centrifuge for 10 s for each wash.
7. Remove the columns from the 1.5 mL Eppendorf tubes, discard the flow-through, and then return the columns to the Eppendorf tubes.
8. Wash columns with 2 × 60 μL of water, centrifuging for 10 s after each addition. Remove the columns from the 1.5 mL Eppendorf tubes, discard the flow-through. Place the columns in labeled new 0.5 mL Eppendorf tubes.

After preparing the cation exchange columns, continue with the following protocol:

9. Pipette solution from the neutralized samples to the corresponding cation exchange columns prepared in advance, centrifuge for 10 s.
10. Wash tubes with 2 × 25 μL Milli-Q water, load onto columns and centrifuge 10 s for each wash.
11. Elute residual glycans from the columns by washing each with 30 μL Milli-Q water and centrifuge for 10 s.

12. Remove cation exchange columns from the Eppendorf tubes, and dry the eluted glycan solutions (remaining in the tubes) to completion in a vacuum centrifuge (45–60 °C). White salt residue (borate) should be visible in each 0.5 mL Eppendorf tube at this stage.
13. Discard the used columns.
14. Remove borate residues by addition of 50 μL desalting solution, and dry to completion in the vacuum centrifuge (45–60 °C).
15. Repeat the addition of desalting solution and subsequent evaporation until no further residues are observed (four to five times) (*see* **Note 6**).
16. Dissolve the released oligosaccharides in Milli-Q water (usually 10–50 μL) for analysis by LC-MS.

3.5 LC-MS of Glycans

The analysis of the released oligosaccharides after incubation with fecal glycosidases requires access to a mass spectrometer capable of CID-MS/MS and an electrospray/nanospray interface. Some examples of LC-MS setups are provided in Table 1. In negative ion mode, oligosaccharides are detected as $[M - nH]^{n-}$-ions. Table 2 provides information about various software and tools that aid in interpretation of data and comparisons of *O*-linked glycan samples analyzed by LC-MS.

1. Perform a data-dependent analysis (DDA) analysis, setting up your LC-MS method to perform a surveyor MS scan followed by two to five data-dependent MS/MS scan.
2. Identify oligosaccharide peaks in your LC-MS chromatogram based on oligosaccharide compositions and/or known fragmentation patterns (Table 2).
3. Export information about the intensity of individual components including intensities of all isotopes and adducts of multiple charge states identified of each component of interest.
4. Perform manual, or semiautomatic assessment of structures in your samples (example in Fig. 2).
5. Assemble MS intensities of individual peaks after digestion together with intensities of equivalent peaks from control samples, and match up substrate and products of sialidases, fucosidases, sulfatases, N-acetylhexosaminidases hexosidases (usually galactosidases).
6. Once peaks are identified and quantified, data analysis such as comparison of individual structures as well as oligosaccharide classes can be applied to extract further information about the glycosidases and their activities in the samples (Table 2) (*see* **Note 7**).

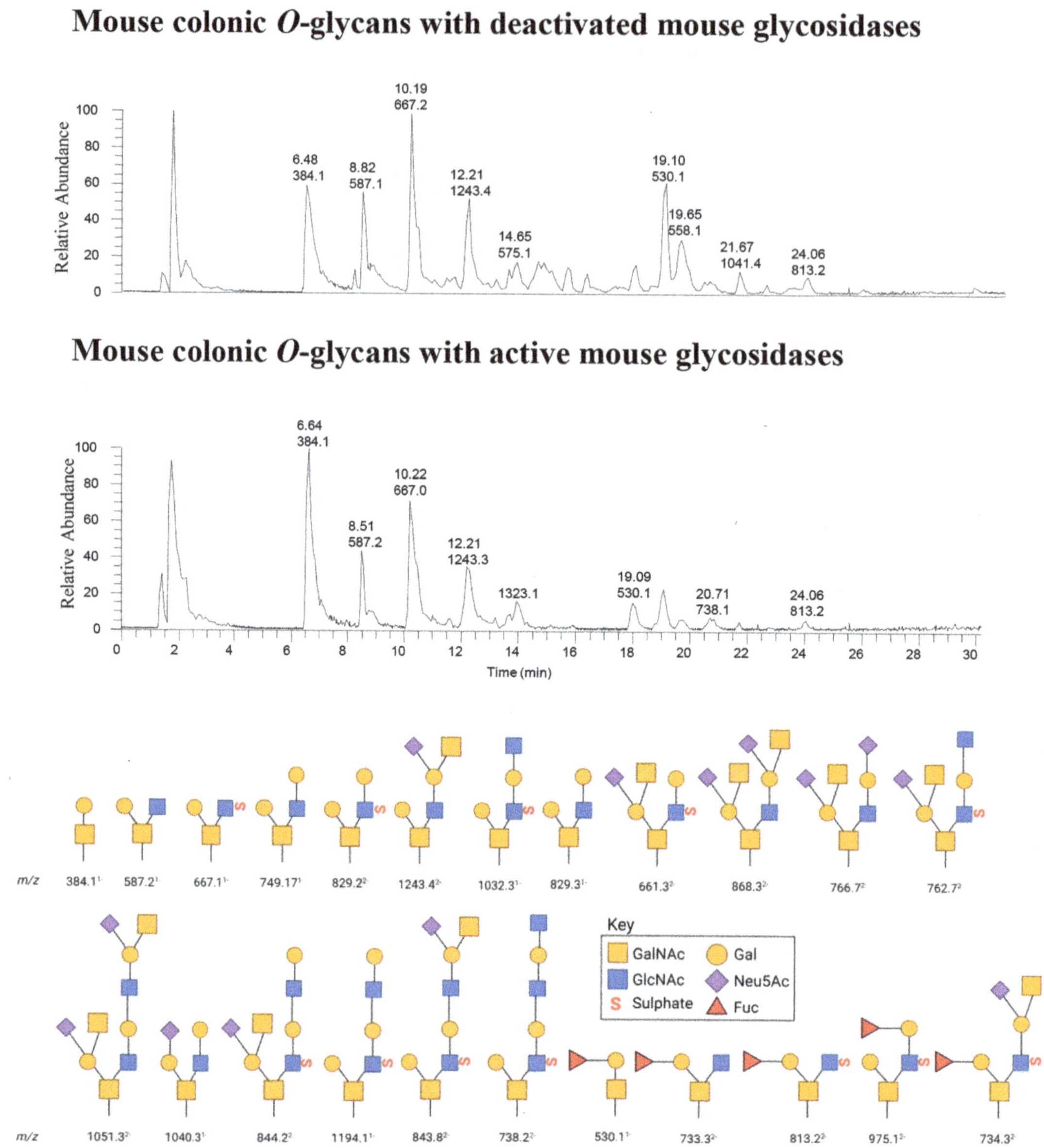

Fig. 2 LC-MS of mouse colonic mucins *O*-glycans incubated with mouse fecal extract. Upper panel shows chromatogram of glycans incubated 2 h with deactivated glycosidases (95 °C). Middle panel shows the LC-MS of glycans after 2 h of incubation. The experiment shows that the relative increase of the $[M - H]^-$- ion of *m/z* 384.1 (Galβ1-3GalNAcol) compared to upper panel is a sign that more complex glycans has been degraded (de-fucosylated) by the extract, for example, the $[M - H]^-$- ion of *m/z* 530.1 (Fucα1-2Galβ1-3GalNAcol). The lower panel shows some examples of oligosaccharides identified from the mouse colon insoluble mucins. Glycan structures are depicted by the SNFG symbols. (Figure partly generated in BioRender (www.biorender.com))

4 Notes

1. Wear gloves and a protective mask while working with PMSF. Highly toxic!
2. The mucins are sometimes difficult to dissolve. If that is the case, dissolve samples into a minimum volume of Extraction GuHCl and dilute with water to make the final concentration 1.0 mg/mL. This solution can be applied directly to the PVDF membrane for dot blotting.
3. Recommended to use several timepoints (e.g. 2, 4, 8, 16 h) to find the optimal incubation time.
4. Always prepare fresh NaOH/$NaBH_4$ solution for each experiment.
5. If any solution has evaporated and condensed on the adhesive coversheet of the microtiter plates during overnight incubation, do not return this solution to the well (results in sugar polymer contamination).
6. These last steps sometimes cause partial methyl esterification of the glycans. If this is a problem, wait to subject the glycans to analysis with LC-MS leaving it in the water solution for 24 h at 4 °C, and the esters are usually hydrolyzed. If this is a reoccurring problem, perform the borate methyl ester evaporation using a desalting solution without acetic acid.
7. To begin the analysis, combine relative intensities of sialylated and non-sialylated, fucosylated, and non-fucosylated as well as sulfated and non-sulfated structures using for instance Excel [12]. Make ratios of these intensities pairs to get an overall view of the type of glycosidase activities present, before making ratios of individual substrate/product intensity pairs.

References

1. Johansson ME, Sjövall H, Hansson GC (2013) The gastrointestinal mucus system in health and disease. Nat Rev Gastroenterol Hepatol 10(6):352–361. https://doi.org/10.1038/nrgastro.2013.35
2. Pickard JM, Zeng MY, Caruso R et al (2017) Gut microbiota: role in pathogen colonization, immune responses, and inflammatory disease. Immunol Rev 279(1):70–89. https://doi.org/10.1111/imr.12567
3. Bergstrom K, Shan X, Casero D et al (2020) Proximal colon–derived O-glycosylated mucus encapsulates and modulates the microbiota. Science 370(6515):467–472. https://doi.org/10.1126/science.aay7367
4. Zhang X, Wang C, Han Q et al (2021) Highly sialylated mucin-type glycopeptide from porcine intestinal mucosa after heparin extraction: O-glycan profiling and immunological activity evaluation. Glycoconj J 38(5):527–537. https://doi.org/10.1007/s10719-021-10014-y
5. Luis AS, Hansson GC (2023) Intestinal mucus and their glycans: a habitat for thriving microbiota. Cell Host Microbe 31(7):1087–1100. https://doi.org/10.1016/j.chom.2023.05.026
6. Wilson NL, Schulz BL, Karlsson NG et al (2002) Sequential analysis of N- and O-linked glycosylation of 2D-PAGE separated

glycoproteins. J Proteome Res 1(6):521–529. https://doi.org/10.1021/pr025538d

7. Sheikh A, Wangdi T, Vickers TJ et al (2022) Enterotoxigenic Escherichia coli degrades the host MUC2 mucin barrier to facilitate critical pathogen-enterocyte interactions in human small intestine. Infect Immun 90(2): e0057221. https://doi.org/10.1128/iai.00572-21
8. Corfield AP (2015) Mucins: a biologically relevant glycan barrier in mucosal protection. Biochim Biophys Acta 1850(1):236–252. https://doi.org/10.1016/j.bbagen.2014.05.003
9. Blöchl C, Wang D, Madunić K et al (2021) Integrated N- and O-glycomics of acute myeloid leukemia (AML) cell lines. Cells 10(11). https://doi.org/10.3390/cells10113058
10. Shathili AM, Brown HM, Everest-Dass AV et al (2018) The effect of streptozotocin-induced hyperglycemia on N-and O-linked protein glycosylation in mouse ovary. Glycobiology 28(11):832–840. https://doi.org/10.1093/glycob/cwy075
11. Hayes CA, Nemes S, Issa S et al (2012) Glycomic work-flow for analysis of mucin O-linked oligosaccharides. Methods Mol Biol 842:141–163. https://doi.org/10.1007/978-1-61779-513-8_8
12. Hayes CA, Nemes S, Karlsson NG (2012) Statistical analysis of glycosylation profiles to compare tissue type and inflammatory disease state. Bioinformatics 28(13):1669–1676. https://doi.org/10.1093/bioinformatics/bts242
13. Damerell D, Ceroni A, Maass K et al (2012) The GlycanBuilder and GlycoWorkbench glycoinformatics tools: updates and new developments. Biol Chem 393(11):1357–1362. https://doi.org/10.1515/hsz-2012-0135
14. Ceroni A, Maass K, Geyer H et al (2008) GlycoWorkbench: a tool for the computer-assisted annotation of mass spectra of glycans. J Proteome Res 7(4):1650–1659. https://doi.org/10.1021/pr7008252
15. Kolarich D, Rapp E, Struwe WB et al (2013) The minimum information required for a glycomics experiment (MIRAGE) project: improving the standards for reporting mass-spectrometry-based glycoanalytic data. Mol Cell Proteomics 12(4):991–995. https://doi.org/10.1074/mcp.O112.026492
16. Rojas-Macias MA, Mariethoz J, Andersso P et al (2019) Towards a standardized bioinformatics infrastructure for N- and O-glycomics. Nat Commun 10(1):3275. https://doi.org/10.1038/s41467-019-11131-x
17. Cooper CA, Gasteiger E, Packer NH (2001) GlycoMod—a software tool for determining glycosylation compositions from mass spectrometric data. Proteomics 1(2):340–349. https://doi.org/10.1002/1615-9861(200102)1:2
18. Watanabe Y, Aoki-Kinoshita KF, Ishihama Y et al (2021) GlycoPOST realizes FAIR principles for glycomics mass spectrometry data. Nucleic Acids Res 49(D1):D1523–D1528. https://doi.org/10.1093/nar/gkaa1012
19. Jin C, Venkatakrishnan V, Thomsson KA et al (2024) UniCarb-DB: an MS/MS experimental glycomic fragmentation database. Methods Mol Biol 2836:77–96. https://doi.org/10.1007/978-1-0716-4007-4_6
20. Ashwood C, Lin CH, Thaysen-Andersen M et al (2018) Discrimination of isomers of released N- and O-glycans using diagnostic product ions in negative ion PGC-LC-ESI-MS/MS. J Am Soc Mass Spectrom 29(6):1194–1209. https://doi.org/10.1007/s13361-018-1932-z
21. Karlsson NG, Schulz BL, Packer NH (2004) Structural determination of neutral O-linked oligosaccharide alditols by negative ion LC-electrospray-MSn. J Am Soc Mass Spectrom 15(5):659–672. https://doi.org/10.1016/j.jasms.2004.01.002
22. Alocci D, Mariethoz J, Gastaldello A et al (2019) GlyConnect: Glycoproteomics goes visual, interactive, and analytical. J Proteome Res 18(2):L664–L677. https://doi.org/10.1021/acs.jproteome.8b00766

Chapter 13

Detection of Bacterial Adhesion to Mucus Slot-Blotted on Polyvinylidene Difluoride (PVDF) Membrane

Macarena P. Quintana-Hayashi and Sara K. Lindén

Abstract

Here, we describe a method that detects bacterial binding to mucus on slot-blot membranes. Mucins, the large, highly glycosylated proteins forming the mucus gel, attach slower to microwell plates than the small less glycosylated components of mucus, which quickly coat plastic surfaces and thereby limit mucin attachment. Therefore, other methods are necessary to retain significant levels of mucins when performing assays on mixed samples. In the method described here, samples are dot-blotted onto a polyvinylidene fluoride membrane and incubated with biotinylated bacteria. Bacterial adhesion to mucins is determined after visualization of the slot-blots with an infrared imaging system and quantification of the image pixel intensities. This method is also useful for the detection of bacterial binding to enzyme-treated mucins, for example with sialidase, and binding intensities can be compared to nonenzyme (mock)-treated mucins.

Key words PVDF membrane, Bacterial adhesion, Bacterial binding, Mucus, Mucins, Slot-blot, Biotinylated bacteria

1 Introduction

Mucus lines the internal epithelia in mammals and in fish also the external surfaces. Mucus contains a multitude of glycans and bacteria often attach to host structures via the glycans [1–5]. The glycans, as well as relative abundance thereof, present on fish skin and intestinal purified mucins and mucus solubilized in chaotropic agent are as similar as technical replicates of purified mucins [6], suggesting that the majority of the glycans in mucus are carried by the mucins. In our experience, the large, highly glycosylated mucins, attach slower to microwell plates commonly used for bacterial binding assays than other small less glycosylated components that can be present in mixed samples such as mucus or mucus extracts. The presence of small molecules that coat plastic surfaces fast in mixed samples can thereby limit mucin attachment. The use of a slot-blot apparatus to suck liquid samples through a

Inka Brockhausen (ed.), *Dynamics of Bacteria-Mucus Interactions*, Methods in Molecular Biology, vol. 2942,
https://doi.org/10.1007/978-1-0716-4627-4_13,

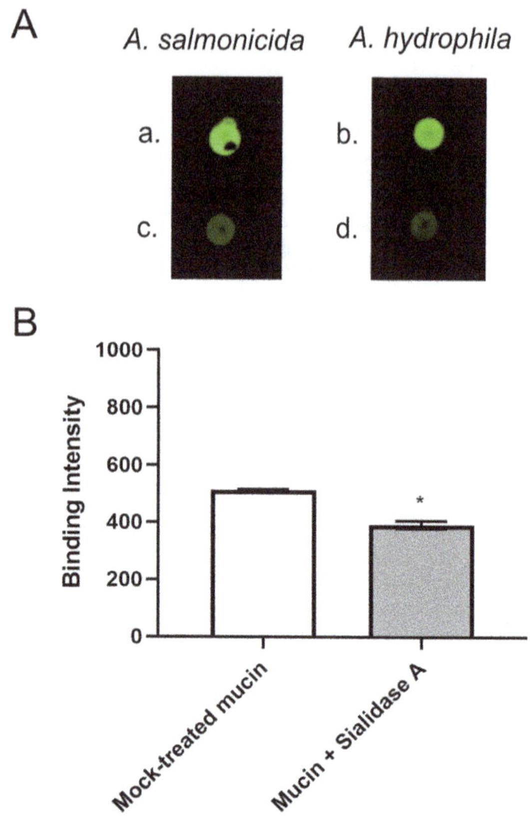

Fig. 1 *A. salmonicida* and *A. hydrophila* binding to mucin containing extracts from Atlantic salmon skin mucus. The mucin containing extracts were prepared as previously described [6]. (**a**) Detection of bacteria binding to Atlantic salmon skin mucin containing extracts dot-blotted onto a PVDF-FL membrane. *A. salmonicida* and *A. hydrophila* binding to a sample containing mucins carrying long glycans (*a.* and *b.* respectively) compared to binding to a sample containing mucins carrying short glycans (*c.* and *d.*). (**b**) *A. salmonicida* binding to sialidase treated mucin containing extracts from Atlantic salmon skin mucus. Binding intensities of *A. salmonicida* to mucin containing extracts from Atlantic salmon skin mucus dot-blotted onto a PVDF membrane and treated with Sialidase A, compared to mock (no enzyme) treated extracts. Results are expressed as the mean ± SEM, *t*-test, $*p < 0.05$

polyvinylidene fluoride (PVDF) membrane allows mucins to attach to the membrane and can therefore be a useful alternative way of assaying bacterial binding to mucin containing samples.

We propose a method for the detection of bacterial adhesion to mucins on slot-blot membranes which we have applied for the detection of *Aeromonas salmonicida* and *Aeromonas hydrophila* binding to fish skin mucin containing mucus extracts (Fig. 1). Using this method, we show a decrease in *A. salmonicida* binding to mucin containing mucus extracts treated with sialidase compared to mock treated extracts (Fig. 1), similar to what we previously have shown for *A. salmonicida* binding to purified Atlantic salmon mucins using micro-titer based assays [7]. Here, we describe its application for the detection of *Aeromonas* spp. binding to Atlantic salmon skin mucins. This method can be modified and applied to a range of bacterial species and mucin containing samples, as we previously have done when studying binding of *Moraxella catarrhalis* to airway samples [3]. Note that some bacterial species carry proteins that bind to biotin. If your samples contain such proteins, it is better to directly label the bacteria with infrared dye instead of biotinylating them, although the sensitivity of the binding assay usually is lower without the biotin-streptavidin amplification step.

2 Materials

1. PVDF-FL membrane (Immobilon FL, Millipore, Bedford, MA).
2. BioDot/BioDot SF Filter Paper (Bio-Rad, CA).
3. Minifold –II Slot-blot system (Schleicher & Schuell Bioscience, Germany).
4. Diluted mucin samples in 6 M GuHCl extraction buffer [6].
5. 100% methanol.
6. Phosphate buffered saline (PBS).
7. Odyssey blocking buffer (LI-COR Biosciences, NE, USA).
8. Carbonate buffer: 0.2 M sodium carbonate pH 8.3.
9. Biotin-XX-NHS: C2OH29N4NaO9S2 (biotinamidocaproic acid 3-sulfo-N-hydroxysuccinimide ester, sodium, Sigma) 125 μg/mL. Dissolve 10 mg in 200 μL dimethylformamide and prepare 20 μL aliquots. Keep frozen.
10. PBS-T, 0.05% Tween 20.
11. PBS-T, 0.2% Tween 20.
12. Secondary antibody: Streptavidin IR dye 800 (Infrared 800 nm, LI-COR, Biosciences, NE, USA).
13. Odyssey infrared imaging system (LI-COR, Biosciences, NE, USA).
14. Enzymes: sialidase A (1 U/200 μL, Prozyme, CA, USA) diluted 1:200, sialidase S (3 U/300 μL Prozyme, CA, USA) diluted 1:200.

3 Methods

Mucin samples are dot-blotted onto a PVDF-FL membrane and incubated with biotinylated bacteria. The slot-blot is then visualized with an infrared imaging system that allows to detect bacterial binding to enzyme-treated and nonenzyme-treated mucins based on intensity.

3.1 Slot-Blot

1. Submerge two BioDot precut absorbent filter papers for blotting applications in diH2O.
2. Cut the PVDF-FL membrane according to the size of the BioDot, and cut off one corner as a marker (*see* **Note 1**).
3. Activate the PVDF-FL membrane by submerging it in methanol for 5 min (*see* **Note 2**).
4. Carefully and using forceps submerge the PVDF-FL membrane in diH2O (*see* **Note 3**).
5. Set up of the Minifold Slot-Blot System in the following order and from bottom up: vacuum manifold, filter support plate, pre-wet filter paper 1, pre-wet filter paper 2, pre-wet PVDF-FL membrane, sample well plate (*see* **Note 4**).
6. Load 100 μL of a diluted mucin sample (5 μg/mL in 6 M GuHCl extraction buffer) per well (*see* **Note 5**).
7. Filter through the membranes by applying a vacuum for approx. 1 h until the mucin samples dry out, starting with a low vacuum and increasing the vacuum as needed (*see* **Note 6**).
8. Once the mucin samples are dry, turn off the vacuum, and carefully disassemble the Minifold Slot-Blot System.
9. Let the blotted PVDF-FL membrane dry for about 1 h (*see* **Note 7**).
10. Once the PVDF-FL membrane is dry, it can be cut according to the layout of the mucin samples loaded and kept inside a clean dry container until bacterial binding assay (*see* **Note 8**).

3.2 Biotinylation of Bacteria

1. Wash freshly harvested bacteria from an agar plate twice in at least 10 mL of carbonate buffer.
2. Centrifuge bacterial suspension at 2500×*g* for 3 min and resuspend in carbonate buffer.
3. Dilute bacterial suspension to obtain approximately 1×10^9 bacteria/mL.
4. Add 2.5 μL of biotin-XX-NHS per 1 mL of bacterial suspension (1×10^9 bacteria/mL), and incubate bacteria for 15 min in the dark, under rotation.
5. Wash bacteria twice in PBS-T (0.05% Tween), centrifuging at 2500×*g* for 3 min each time.
6. Dilute biotinylated bacteria 1:10 in Odyssey blocking buffer containing 0.1% Tween 20.

3.3 Bacterial Binding to Mucins on PVDF-FL Membrane

1. Pre-wet the slot-blotted membrane with 30 mL 100% methanol for 5 min until the membrane sinks (*see* **Note 9**).
2. Rinse membrane with 30 mL diH2O, and incubate the membrane in 30 mL PBS for 10 min under orbital shaking (*see* **Note 10**).
3. Block the membrane in 30 mL Odyssey blocking buffer for 1 h at RT under orbital shaking.
4. Incubate the membrane with biotinylated bacteria (refer to **step 6**, Subheading 3.2) further diluted 1:20 in Odyssey blocking buffer containing 0.1% Tween 20 for 2 h at RT with gentle orbital shaking.
5. Wash the membrane three times for 5 min each with PBS-T (0.2% Tween) with gentle shaking.
6. Incubate the membrane with the secondary antibody diluted 1: 10000 in Odyssey blocking buffer containing 0.1% Tween 20 for 30 min in the dark with gentle shaking.
7. Wash the membrane three times for 5 min with PBS-T (0.2% Tween) and a final rinse with PBS.
8. Image with the infrared imaging system and quantify the bacterial binding intensity with the analysis software provided (*see* **Note 11**).

3.4 Bacterial Binding to Enzyme-Treated Mucins on PVDF-FL Membrane

1. Proceed to **steps 1–10**, Subheading 3.1, **1–6**, Subheading 3.2, and **1–3**, Subheading 3.3.
2. Incubate the mucins slot-blotted onto the PVDF-FL membrane with 500 μL of enzyme (e.g., sialidase) per well at 37 ° C overnight (*see* **Note 12**).
3. Rinse membrane twice with PBS.
4. Proceed to **step 3–8**, Subheading 3.3.

4 Notes

1. Preferably the PVDF-FL membrane should be cut with its blue peel on and on top of a cutting board using a scalpel.
2. The PVDF-FL membrane should be submerged in methanol with its blue peel on. The blue peel should be carefully removed after activation using small forceps. Alternatively, dried methanol can be used for activation of the PVDF-FL membrane.
3. In all steps, the PVDF-FL membrane should be handled with care from the corner and using a small forceps, as touching and any tweezer marks will fluoresce with the infrared imaging system.

4. The manifold system should be screwed tightly, preferably screwing opposite screws simultaneously to avoid mucin sample leakage.
5. Mucin samples can be loaded in duplicates. Mucin samples should be loaded fast before the PVDF-FL membrane dries out and starting from the side closest to the vacuum.
6. If there are wells with mucin samples that have not completely dried out after 1 h, carefully pipette the mucin sample up and down to remove any bubbles without touching the membrane.
7. To let the PVDF-FL membrane dry, it can be placed on top of a BioDot filter paper and inside a clean box covered with a lid or foil.
8. To cut the PVDF-FL membrane, place it on top of a clean piece of glass with a BioDot filter paper (with marked lines as a guide) under it. Cut the PVDF-FL membrane using a scalpel or scissors, and cut the bottom right corner of each membrane piece to mark the position of the bottom mucin samples. Keep the PVDF-FL membrane in a clean and dry covered container (e.g., petri dish).
9. The 30 mL of methanol, PBS, and blocking buffer stated are only for reference, this volume will depend on the size of the container used and should be enough to completely cover the membrane.
10. The containers used for submerging, washing, and incubating the PVDF-FL membranes should be free from blue staining residues as blue stain gives background in the image.
11. To prevent the membrane from drying out, do not remove it from PBS until imaging.
12. To reduce the total volume of enzyme needed to treat the mucins slot-blotted on a PVDF-FL membrane, it is recommended to cut individual slot-blots and place each one inside a 12-well plate for enzyme treatment. Nonenzyme (mock)-treated mucins should be included as controls.

Acknowledgments

This work was supported by the Swedish Research Council Formas (2018-01419), the heart-lung foundation (20230626), and the Carl Trygger Foundation (CTS 23:2971).

References

1. Lindén S et al (2002) Strain- and blood group-dependent binding of helicobacter pylori to human gastric MUC5AC glycoforms. Gastroenterology 123(6):1923–1930

2. Padra M et al (2019) Helicobacter suis infection alters glycosylation and decreases the pathogen growth inhibiting effect and binding avidity of gastric mucins. Mucosal Immunol 12(3): 784–794
3. Padra M et al (2021) Mucin binding to Moraxella catarrhalis during airway inflammation is dependent on sialic acid. Am J Respir Cell Mol Biol 65(6):593–602
4. Quintana-Hayashi MP et al (2019) Role of sialic acid in Brachyspira hyodysenteriae adhesion to pig colonic mucins. Infect Immun 87(7): e00889–e00818
5. Thomsson KA et al (2022) Mucin O-glycosylation and pathogen binding ability differ between rainbow trout epithelial sites. Fish Shellfish Immunol 131:349–357
6. Benktander J et al (2019) Effects of size and geographical origin on Atlantic salmon, Salmo salar, mucin O-glycan repertoire. Mol Cell Proteomics 18(6):1183–1196
7. Padra JT et al (2014) Aeromonas salmonicida binds differentially to mucins isolated from skin and intestinal regions of Atlantic salmon in an N-acetylneuraminic acid-dependent manner. Infect Immun 82(12):5235–5245

Chapter 14

Quantification of β-Lactamase Inhibition Using a Luminescent Whole-Cell Biosensor

Mitchell A. Jeffs and Christopher T. Lohans

Abstract

The clinical utility of the β-lactam antibiotics is threatened by resistant bacterial pathogens that produce β-lactamases. These enzymes hydrolytically degrade β-lactams, rendering them ineffective. Although several β-lactamase inhibitors have been approved for clinical use, currently used inhibitors only target a subset of the clinically relevant β-lactamases. Additionally, β-lactamase variants that exhibit reduced susceptibility to these inhibitors are increasing in prevalence. Thus, drug discovery efforts must continue to identify and develop novel β-lactamase inhibitors. Herein, we describe a luminescent biosensor, based on the β-lactam-inducible AmpR regulon, which can be used to quantify β-lactamase inhibition in bacterial cells. This approach does not require the use of expensive substrates and can account for factors such as outer membrane permeability, prioritizing the discovery of inhibitors that are effective against bacterial cells.

Key words Antimicrobial resistance, Biosensor, β-Lactamase inhibitors, Drug discovery, Reporter assay

1 Introduction

β-Lactams are currently the most frequently prescribed antibiotics globally [1]. Members of this drug class include the penicillins, cephalosporins, carbapenems, and monobactams. Many β-lactams exhibit a broad spectrum of antimicrobial activity, targeting both Gram-positive and Gram-negative pathogens. However, the clinical utility of these antibiotics is endangered by the global spread of β-lactamase-producing bacteria [2]. β-lactamase enzymes inactivate β-lactams by hydrolyzing the core β-lactam ring of the antibiotic. Consequently, the degraded antibiotics are no longer able to inhibit penicillin-binding proteins, their primary targets in the bacterial cell. β-lactamases are broadly classified based on their hydrolytic mechanism; serine β-lactamases (SBLs) utilize a nucleophilic serine residue, while metallo-β-lactamases (MBLs) employ one or two zinc ions to catalyze β-lactam hydrolysis. These enzymes are

Inka Brockhausen (ed.), *Dynamics of Bacteria-Mucus Interactions*, Methods in Molecular Biology, vol. 2942, https://doi.org/10.1007/978-1-0716-4627-4_14,

commonly further broken down according to the Ambler classification system: β-lactamases belonging to classes A, C, and D are SBLs, while class B enzymes are MBLs. β-lactamases can also be classified based on the types of β-lactam antibiotics that they are capable of degrading. Penicillinases (e.g., TEM-1, SHV-1) hydrolyze penicillins and some early-generation cephalosporins (e.g., cefazolin, cephalothin), while extended-spectrum β-lactamases (ESBLs; e.g., CTX-M-15, some TEM and SHV variants) also inactivate later generation cephalosporins (e.g., ceftriaxone, cefotaxime). Of particular concern to global health are the carbapenemases (e.g., OXA-48, KPC-2, NDM-1, IMP-1, VIM-2) due to their ability to hydrolyze carbapenems, antibiotics of last resort [3–5].

To overcome β-lactamase-mediated resistance, several β-lactamase inhibitors (BLIs) are currently employed in the clinic. These drugs are administered alongside a partner β-lactam, protecting the antibiotic from β-lactamase-catalyzed hydrolysis. The first widely used BLIs, clavulanic acid, sulbactam, and tazobactam, are β-lactam-based drugs with a limited spectrum of inhibition, mainly targeting penicillinases and some ESBLs [6, 7]. In 2015, the non--β-lactam-based BLI avibactam was approved for clinical use in combination with the cephalosporin ceftazidime. Avibactam exhibits potent activity against class A and C SBLs (including carbapenemases such as KPC-2), but it is less potent against certain class D SBLs and does not inhibit MBLs [8–10]. Additionally, resistance to ceftazidime-avibactam was reported less than 2 years after its approval for clinical use; since this time, several resistance mechanisms against this drug combination have been identified [11–14]. Two other novel β-lactam-BLI combinations (imipenem-relebactam and meropenem-vaborbactam) have also been recently approved for clinical use. However, like avibactam, the BLIs relebactam and vaborbactam do not inhibit MBLs [15, 16]. Currently, the only MBL inhibitor nearing clinical approval is taniborbactam, which recently completed a phase III clinical trial where it was administered in combination with the cephalosporin cefepime. Taniborbactam exhibits broad-spectrum activity against all Ambler classes of β-lactamases, including some MBLs such as NDM-1 and VIM-2 [17]. Although the combination of cefepime-taniborbactam shows significant clinical promise, it is less potent against certain MBLs including IMP-type enzymes, and resistance to this drug combination has already been reported [18, 19]. Thus, further discovery efforts are needed to advance the identification and development of new broad spectrum BLIs, particularly those that are active against MBLs.

Most BLI screening campaigns employ in vitro assays in which compound libraries are tested against purified recombinant β-lactamases. One widely used approach uses the chromogenic cephalosporin nitrocefin, which undergoes a color change when it is hydrolyzed by a β-lactamase. Other chromogenic and fluorogenic β-lactam substrates have also been reported for testing inhibitors against purified β-lactamase enzymes [20, 21]. However, the use of these substrates is limited by the costs associated with their synthesis or acquisition. Additionally, screening methods which test BLIs against purified enzymes do not account for cellular factors that can impact inhibitor potency. Traditionally, broth microdilution-based minimum inhibitory concentration (MIC) testing is used to assess the efficacy of β-lactam-BLI combinations against bacterial cells. One key drawback of this MIC testing is the lack of specificity; reductions in cell growth do not specifically indicate β-lactamase inhibition and can instead result from the antimicrobial activity of a BLI candidate. Furthermore, it is challenging to use MIC testing to quantitatively assess the inhibitory potency of a BLI. Whole-cell assays employing UV-Vis spectrophotometry to monitor β-lactamase inhibition have been developed as an alternative strategy [22]. However, these methods can suffer from poor sensitivity when used to study weaker β-lactamases, and spectral overlap of potential inhibitors with assay substrates can interfere with measurements of β-lactamase activity.

To address the limitations associated with current testing methods, we developed a luminescent whole-cell biosensor for the quantification of β-lactamase inhibition [23]. The biosensor employs the AmpR/AmpC regulatory system found in several bacteria species, including *Pseudomonas aeruginosa*. In *P. aeruginosa*, exposure to certain β-lactam antibiotics leads to the increased production of 1,6-anhydromuropeptide peptidoglycan catabolites [24]. These catabolites bind to the AmpR transcription factor, resulting in the de-repression of the *ampC* promoter and the production of the β-lactamase AmpC. We prepared the plasmid pAMPLUX in which expression of the *luxCDABE* luciferase reporter is regulated by the *ampC* promoter and AmpR. While the exposure of biosensor cells (i.e., *Escherichia coli* BW25113 transformed with the plasmid pAMPLUX) to β-lactam antibiotics results in a luminescent signal, the addition of a β-lactamase-producing strain to the mixture results in antibiotic degradation and reduced luminescence. Thus, addition of an effective BLI rescues the β-lactam and restores the luminescent signal. This assay is simple to set up, rapid, and does not require the use of potentially costly substrates. Furthermore, unlike assays with purified β-lactamase enzymes, the biosensor assay is selective for inhibitors that target β-lactamases in a cellular context. This allows for other factors which can impact BLI efficacy to be accounted for, such as cell wall permeability and efflux [25–27].

2 Materials

2.1 Required Materials for Bacterial Culturing and Biosensor Assay

1. 2× tryptone-yeast extract media, 2TY: 16 g/L tryptone, 10 g/L yeast extract, 5 g/L NaCl or cation-adjusted Mueller-Hinton Broth, CAMHB.
2. 2TY or CAMHB agar plates supplemented with (a) 50 μg/mL kanamycin for biosensor cell selection, or (b) an appropriate antibiotic to select for bacteria transformed with plasmids containing β-lactamase genes. In our study, 25 μg/mL chloramphenicol was used to select for bacteria transformed with β-lactamase-encoding pACYC184 plasmids.
3. Whole-cell biosensor: *E. coli* BW25113 was transformed with the pAMPLUX plasmid [23].
4. Negative control strain: any non-β-lactamase-producing *E. coli* strain (e.g., K-12, BW25113, ATCC 25922, etc.).
5. β-lactamase-producing bacterial strain(s): in our study, a panel of β-lactamase-producing bacterial strains was generated by transforming *E. coli* BW25113 cells with pACYC184 plasmids encoding different β-lactamases (TEM-116, KPC-2, OXA-48, NDM-1, IMP-1 and VIM-2).

2.2 Reagents and Solvents

1. Antibiotics used for selection purposes during bacterial culturing: ampicillin, kanamycin, and chloramphenicol.
2. Antibiotics for biosensor assays: imipenem or meropenem for assays using carbapenemases, amoxicillin for assays using penicillinases or extended-spectrum β-lactamases.
3. β-lactamase inhibitor(s) of interest.
4. Sterile water (or another appropriate solvent) to dissolve antibiotics and β-lactamase inhibitors (*see* **Note 1**).

2.3 General Supplies

1. Sterile Petri dishes.
2. Sterile inoculation loops.
3. Sterile 1.5 mL, 15 mL, or 50 mL plastic tubes: the types of tubes needed will depend on the number of inhibitors and bacterial strains being tested.
4. Cuvettes.
5. White 96-well microplates (Greiner Bio-One LUMITRAC 200 or equivalent).
6. Spectrophotometer for OD_{600} measurements.
7. Plate reader capable of luminescence measurements.

3 Methods

For a graphical overview of the methodology described below, see Fig. 1. The protocol outlined in this methods section describes a dose-response analysis experiment for determining inhibitor potency against β-lactamase-producing bacterial cells. Note that this protocol can also be adapted to "single-shot" primary screening of unknown compounds for inhibitory activity against β-lactamases (*see* **Note 2**).

3.1 Day Prior to Experiment

Starting from a frozen glycerol stock, use a sterile inoculation loop to prepare streak plates of biosensor cells (*E. coli* BW25113 transformed with pAMPLUX) on 2TY or CAMHB agar plates supplemented with 50 μg/mL kanamycin. Streak plates of bacteria producing the β-lactamases of interest, as well as a negative control strain (i.e., a non-β-lactamase-producer), should also be prepared on 2TY or CAMHB agar plates supplemented with antibiotic if

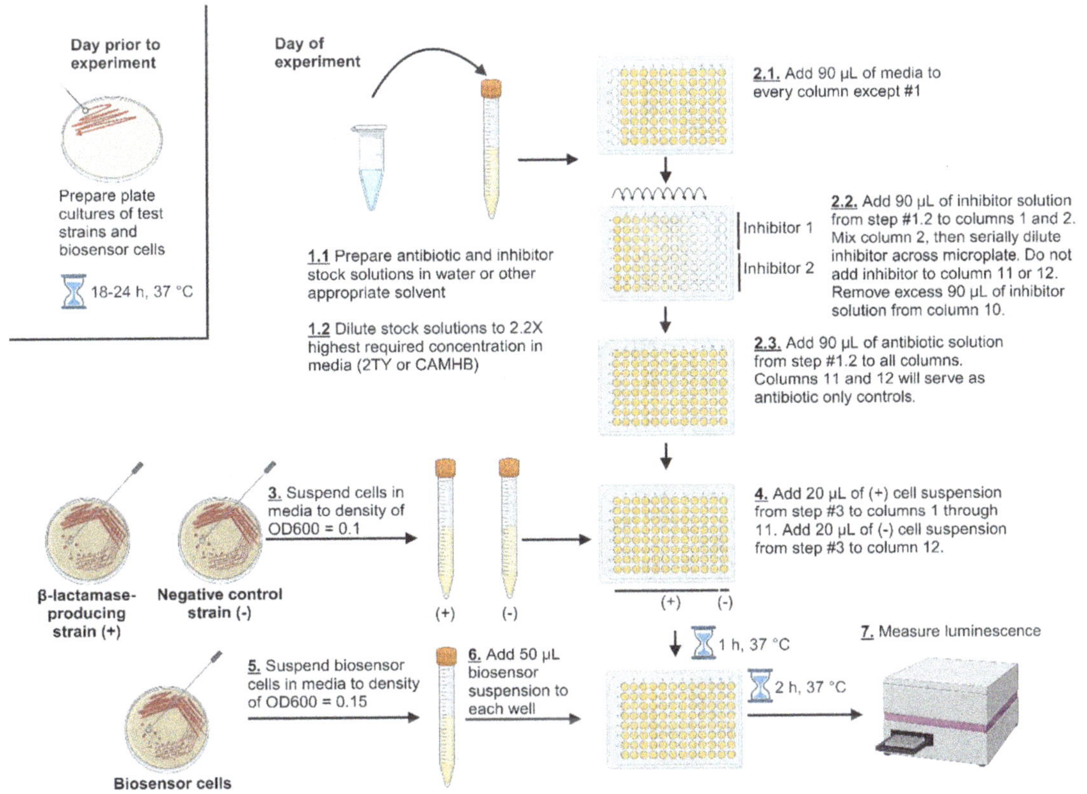

Fig. 1 Graphical overview of the biosensor assay protocol. Note the number of microplate rows utilized in the assay will depend on the number of inhibitors and replicates being tested. The number of columns may differ depending on how many inhibitor concentrations are assayed. This figure outlines the testing of two inhibitors at 10 different concentrations with four technical replicates. Created in BioRender. Jeffs, M. (2025) https:/BioRender.com/hfxthqb

required. Cultures should be grown overnight (16–20 h) at 37 °C without shaking. As indicated above, culturing and assay experiments can use either CAMHB or 2TY; for the purpose of brevity, the experimental protocol provided below will only mention CAMHB.

3.2 Experimental Protocol on Day of Experiment

1. Prepare the required β-lactam and β-lactamase inhibitor stock solutions in the appropriate solvents (*see* **Note 3**).
2. Prepare working solutions of each antibiotic and inhibitor in sterile CAMHB in 1.5 mL, 15 mL or 50 mL tubes, depending on the required volume for the assay. The concentrations of these solutions should be 2.2X greater than the highest concentration required for the assay, so as to account for dilutions that will be carried out in the microplate. For example, if an antibiotic concentration of 2 μg/mL is required in the assay, the concentration of the working solution should be 4.4 μg/mL. Penicillins are weaker inducers of the biosensor, resulting in a less intense luminescent signal compared to carbapenems. As such, carbapenems (e.g., imipenem, meropenem) should be prepared to a concentration of 2.2 μg/mL, while penicillins (e.g., amoxicillin) should be prepared to a concentration of 17.6 μg/mL.
3. To a white 96-well microplate (*see* **Note 4**), add 90 μL of sterile CAMHB to all of the wells that will be used in columns 2–12; note that the number of columns used may vary depending on how many inhibitor concentrations are being tested. Two columns should be reserved for controls which will not contain inhibitor. At least three technical replicates should be prepared for all treatments, and so at least three rows will be needed for each inhibitor tested.
4. Add 90 μL of the inhibitor solution(s) prepared in step 2 to columns 1 and 2 of the white microplate. Mix the contents of column 2 by pipetting up and down, then prepare a twofold serial dilution by transferring 90 μL of solution from column 2 to column 3 and mix by pipetting. Repeat this process for the subsequent columns of the microplate. Do not transfer any inhibitor into the last two columns. Note the final column receiving the serially diluted inhibitor will have twice the required volume. Therefore, the excess 90 μL from this column can be discarded, leaving a final volume of 90 μL in these wells.
5. After serial dilutions of the inhibitor(s) have been prepared in the microplate, add 90 μL of the antibiotic solution (prepared in step 2) to all of the wells that contain inhibitors, as well as to the last two columns of the microplate which will contain antibiotic alone.

6. Prepare suspensions of the β-lactamase-producing bacterial strains and the negative control strain by using a sterile inoculation loop to suspend colonies from the streak plates in sterile CAMHB in 15 mL tubes. Ensure the suspensions are well-mixed and that any cell clumps are broken up by vortexing for 5–10 s. The volume of cell suspensions required will depend on how many inhibitors are being tested. Ensure that a large enough volume is prepared so that each strain can be tested with at least triplicate measurements for every inhibitor concentration, based on the total volume of cells added to the microplate in **step** 7. Also ensure there is sufficient excess volume so that at least 1 mL remains for OD_{600} readings in a cuvette. Typically, suspensions should be prepared to an $OD_{600} = 0.1$, as determined with a spectrophotometer (*see* **Note 5,** Fig. 2).
7. Add 20 μL of the suspension of β-lactamase-producing cells to columns 1 through 11 of the white microplate. Then, add 20 μL of the suspension of the negative control strain to column 12 of the microplate. Incubate the plate at 37 °C for 1 h (*see* **Note 6**).
8. After approx. 50 min of incubation, begin preparing a suspension of biosensor cells in sterile CAMHB to an OD_{600} of 0.15. The total volume needed will depend on the number of

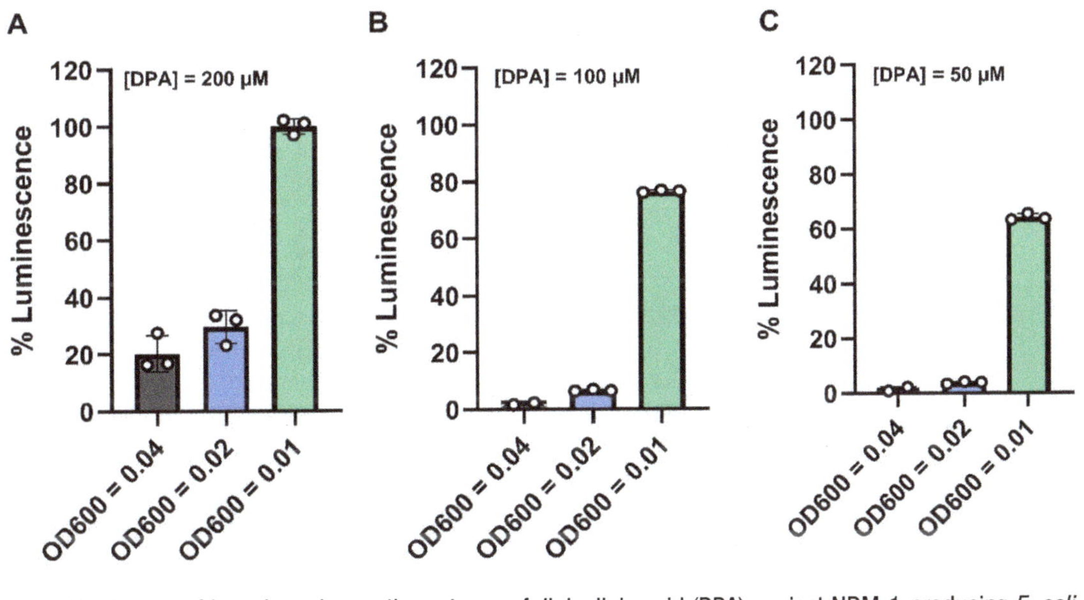

Fig. 2 The impact of inoculum size on the potency of dipicolinic acid (DPA) against NDM-1-producing *E. coli*. Suspensions of NDM-1-producing *E. coli* were prepared to $OD_{600} = 0.4$, 0.2 or 0.1, then diluted tenfold in 2TY supplemented with meropenem (final concentration 1 μg/mL) and the indicated concentration of DPA. Luminescence readings were normalized to the sample with the greatest luminescence. $n = 3$, error bars indicate S.D. (Figure adapted from Jeffs et al. [23])

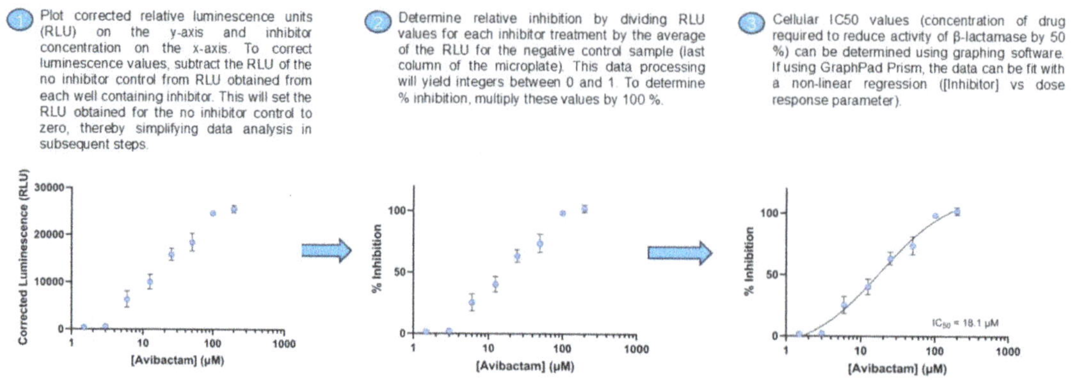

Fig. 3 Overview of data processing and analysis. The steps outlined in the figure describe how the cellular IC_{50} for a β-lactamase inhibitor against a specific β-lactamase is determined based on the relative luminescence units (RLU) measurements obtained from the luminometer. This example illustrates data obtained from an inhibition assay measuring the efficacy of the β-lactamase inhibitor avibactam against the β-lactamase KPC-2 in an assay using the antibiotic meropenem

inhibitors and β-lactamase-producing strains being tested. Make sure to prepare a sufficient volume so that each condition can be tested in triplicate (*see* **step 9**), while also reserving a sufficient volume for OD_{600} measurements.

9. Once the 1 h incubation in **step 7** is complete, add 50 μL of the biosensor cell suspension to all of the wells in the white microplate. Incubate the microplate at 37 °C and take luminescence readings after 2 h.

3.3 Data Analysis

Representative data analysis steps that follow acquisition of luminescence readings are shown in Fig. 3. In short, the average luminescence values from the microplate wells which contain β-lactamase-producing strain(s) that were not treated with inhibitor (which should yield low luminescence values) should be subtracted from the readings of all of the microplate wells which did receive inhibitor. This blanks the luminescence in the no-inhibitor control to 0, thus serving as a baseline for inhibition. All luminescence values should then be normalized to the luminescence readings obtained for the negative control strain (which should yield high luminescence values).

4 Notes

1. Organic solvents can be used to dissolve drug powders if needed; however, these stocks should be sufficiently diluted in assay media (CAMHB or 2TY) to minimize the potential impact of the solvent on the biosensor cells. Test experiments

evaluating the effect of diluted solvent on luminescence production by biosensor cells in the absence of antibiotic are highly recommended.

2. Instead of preparing serial dilutions of the inhibitor in the microplate, as indicated in Subheading 3.2, a single inhibitor concentration can be used. In short, 90 μL of each compound would be added to microplate wells in triplicate, and then these wells would be supplemented with 90 μL of antibiotic solution. Note that the control solution with antibiotic alone is still required, as described in Subheading 3.2.
3. Antibiotic and inhibitor stocks should be prepared on the day of the experiment, particularly if they are known to be unstable. Note that organic solvents such DMSO may be used to dissolve drug powders if needed; however, these stocks will first need to be diluted in assay media (CAMHB or 2TY) to minimize the potential impact on the biosensor cells. It is highly recommended to evaluate the effect of the diluted solvent on the biosensor cells in the absence of inhibitors and antibiotic prior to running an assay. If a major reduction in luminescence is observed at the solvent concentrations tested, the solvent will need to be further diluted.
4. This assay protocol was carried out using Greiner Bio-One LUMITRAC 200 white 96-well microplates and a Synergy LX multimode plate reader (Agilent BioTek) with luminescence gain set to 130 (default setting). While other plates and plate readers can be used in this assay, some parameters (e.g., gain) may need to be adjusted.
5. The density of the β-lactamase-producing cell suspension can be adjusted to optimize the assay for the identification of stronger or weaker inhibitors during primary screening experiments. β-lactamase production can contribute to the inoculum effect phenomenon, where the efficacy of β-lactam antibiotics and β-lactamase inhibitors is reduced at a higher inoculum of bacteria [28]. Figure 2 illustrates the impact of inoculum size on the inhibitory potency of dipicolinic acid against NDM-1-producing *E. coli* cells. Greater cell densities lead to increased amounts of β-lactamase in the assay media, thus requiring higher concentrations of inhibitor to prevent β-lactam hydrolysis.
6. Incomplete hydrolysis of the antibiotic may occur if the assay is carried out with strains of bacteria that produce only low levels of β-lactamases, or for β-lactamases with weak hydrolytic activity toward the target antibiotic. The extent of hydrolysis can be determined by comparing the luminescence intensity of the column which contains only antibiotic and the β-lactamase-producing strain (column 11 in the protocol described above)

to that of the negative control (column 12). Based on our experience, if the luminescence of column 11 is more than half that of column 12, this suggests incomplete hydrolysis. If this occurs, the length of preincubation with the target antibiotic can be increased (e.g., to 1.5 or 2 h), or the inoculum of β-lactamase-producing cells can be increased.

References

1. Bush K, Bradford PA (2016)-lactams and β-lactamase inhibitors: an overview. Cold Spring Harb Perspect Med 6:a025247
2. Bush K, Bradford PA (2020) Epidemiology of β-lactamase-producing pathogens. Clin Microbiol Rev 33. https://doi.org/10.1128/cmr.00047-19
3. Bonomo RA, Burd EM, Conly J et al (2018) Carbapenemase-producing organisms: a global scourge. Clin Infect Dis 66:1290–1297
4. Pournaras S, Zarrilli R, Higgins PG et al (2021) Editorial: carbapenemase-producing organisms as leading cause of hospital infections. Front Med 8:775021
5. Van Duin D, Doi Y (2017) The global epidemiology of carbapenemase-producing enterobacteriaceae. Virulence 8:460–469
6. Monogue ML, Heil EL, Aitken SL et al (2021) The role of tazobactam-based combinations for the management of infections due to extended-spectrum β-lactamase-producing enterobacterales: insights from the Society of Infectious Diseases Pharmacists. Pharmacotherapy 41:864–880
7. Payne DJ, Cramp R, Winstanley DJ et al (1994) Comparative activities of clavulanic acid, sulbactam, and tazobactam against clinically important beta-lactamases. Antimicrob Agents Chemother 38:767–772
8. Wang DY, Abboud MI, Markoulides MS et al (2016) The road to avibactam: the first clinically useful non-β-lactam working somewhat like a β-lactam. Future Med Chem 8:1063–1084
9. Ehmann DE, Jahić H, Ross PL et al (2013) Kinetics of avibactam inhibition against class A, C, and D β-lactamases. J Biol Chem 288: 27960–27971
10. Ehmann DE, Jahić H, Ross PL et al (2012) Avibactam is a covalent, reversible, non–-β-lactam β-lactamase inhibitor. Proc Natl Acad Sci 109:11663–11668
11. Di Bella S, Giacobbe DR, Maraolo AE et al (2021) Resistance to ceftazidime/avibactam in infections and colonisations by KPC-producing enterobacterales: a systematic review of observational clinical studies. J Glob Antimicrob Resist 25:268–281
12. Li X, Quan J, Ke H et al (2021) Emergence of a KPC variant conferring resistance to ceftazidime-avibactam in a widespread ST11 carbapenem-resistant *Klebsiella pneumoniae* clone in China. Front Microbiol 12:724272
13. Wang Y, Wang J, Wang R et al (2020) Resistance to ceftazidime–avibactam and underlying mechanisms. J Glob Antimicrob Resist 22:18–27
14. Shields RK, Chen L, Cheng S et al (2017) Emergence of Ceftazidime-avibactam resistance due to plasmid-borne blaKPC-3 mutations during treatment of carbapenem-resistant *Klebsiella pneumoniae* infections. Antimicrob Agents Chemother 61. https://doi.org/10.1128/aac.02097-16
15. Mansour H, Ouweini AEL, Chahine EB et al (2021) Imipenem/cilastatin/relebactam: a new carbapenem β-lactamase inhibitor combination. Am J Health Syst Pharm 78:674–683
16. Lomovskaya O, Sun D, Rubio-Aparicio D et al (2017) Vaborbactam: spectrum of beta-lactamase inhibition and impact of resistance mechanisms on activity in enterobacteriaceae. Antimicrob Agents Chemother 61:e01443–e01417
17. Hamrick JC, Docquier JD, Uehara T et al (2020) VNRX-5133 (taniborbactam), a broad-spectrum inhibitor of serine- and metallo-β-lactamases, restores activity of cefepime in enterobacterales and *Pseudomonas aeruginosa*. Antimicrob Agents Chemother 64:e01963–e01919
18. Drusin SI, Le Terrier C, Poirel L et al (2023) Structural basis of metallo-β-lactamase resistance to taniborbactam. Antimicrob Agents Chemother 68:e01168–e01123
19. Karlowsky JA, Hackel MA, Wise MG et al (2023) *In vitro* activity of cefepime-taniborbactam and comparators against clinical isolates of gram-negative bacilli from 2018 to 2020: results from the global evaluation of antimicrobial resistance via surveillance

(GEARS) program. Antimicrob Agents Chemother 67:e01281–e01222

20. Bebrone C, Moali C, Mahy F et al (2001) CENTA as a chromogenic substrate for studying beta-lactamases. Antimicrob Agents Chemother 45:1868–1871
21. van Berkel SS, Brem J, Rydzik AM et al (2013) Assay platform for clinically relevant metallo-β-lactamases. J Med Chem 56:6945–6953
22. Ge Y, Zhou YJ, Yang KW et al (2017) Real-time activity assays of β-lactamases in living bacterial cells: application to the inhibition of antibiotic-resistant *E. coli* strains. Mol BioSyst 13:2323–2327
23. Jeffs MA, Gray RAV, Sheth PM et al (2023) Development of a whole-cell biosensor for β-lactamase inhibitor discovery. Chem Commun 59:12707–12710
24. Torrens G, Hernández SB, Ayala JA et al (2019) Regulation of AmpC-driven β-lactam resistance in *Pseudomonas aeruginosa*: different pathways, different signaling. mSystems 4: e00524–e00519
25. Ghai I (2023) A barrier to entry: examining the bacterial outer membrane and antibiotic resistance. Appl Sci 13. https://doi.org/10.3390/app13074238
26. Nelson K, Hemarajata P, Sun D et al (2017) Resistance to ceftazidime-avibactam is due to transposition of KPC in a porin-deficient strain of *Klebsiella pneumoniae* with increased efflux activity. Antimicrob Agents Chemother 61. https://doi.org/10.1128/aac.00989-17
27. Chalhoub H, Sáenz Y, Nichols WW et al (2018) Loss of activity of ceftazidime-avibactam due to MexAB-OprM efflux and overproduction of AmpC cephalosporinase in *Pseudomonas aeruginosa* isolated from patients suffering from cystic fibrosis. Int J Antimicrob Agents 52:697–701
28. Danjean M, Hobson CA, Gits-Muselli M et al (2022) Evaluation of the inoculum effect of new antibiotics against carbapenem-resistant Enterobacterales. Clin Microbiol Infect 28: 1503.e1–1503.e3

Chapter 15

Mucins and Bacterial Interactions

Yannick Rossez, Renaud Léonard, Olivier Vidal, Zina Chouit, Yann Guerardel, Edwige Madec, and Catherine Robbe-Masselot

Abstract

The handling of high-molecular-weight mucins requires specialized methods. The present manuscript introduces a slot blotting-based technique to screen post-ultracentrifugation fractions using colorimetric assays. Next, we describe an easy slot blot-based method to characterize mucin-bacteria interactions and to assess the effects of glycan enzymatic treatment on bacterial adhesion.

Key words Mucin, Adhesion, Glycans, Bacteria, Slot blot

1 Introduction

The intestinal mucosa serves as a primary site for colonization by various pathogens, while also secreting a protective layer of thick mucus that shields the underlying epithelium [1–3] . This protective function is largely attributed to the highly *O*-glycosylated gel-forming mucins that make up the mucus layer [4–6]. Due to their high molecular weight and physicochemical properties, handling and purifying mucins is challenging and requires dedicated protocols [7–9]. An effective approach to follow-up mucin purification is to use slot blotting to immobilize purified samples on nitrocellulose or polyvinylidene difluoride membranes. Fractions can then be selected after ultracentrifugation using colorimetric assays such as lectin staining or immunodetection to identify and isolate the desired mucin fractions. Once purified, slot blotting can be adapted to study mucin-bacteria interactions [10]. In this study, we describe two staining methods to identify mucin-containing fractions after purification by ultracentrifugation (Fig. 1). The purified mucins were then used to study their interactions with bacteria (Figs. 2 and 3). To investigate the role of sialic acid in bacterial adhesion, the mucins were subjected to enzymatic treatments. The ability of bacteria to adhere to these treated mucins was then

Inka Brockhausen (ed.), *Dynamics of Bacteria-Mucus Interactions*, Methods in Molecular Biology, vol. 2942,
https://doi.org/10.1007/978-1-0716-4627-4_15,

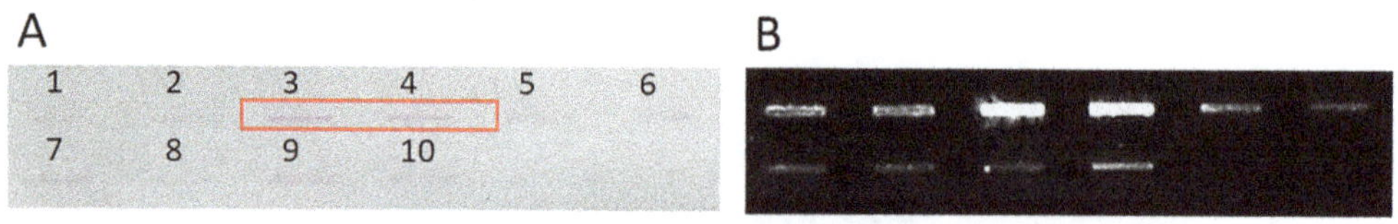

Fig. 1 Slot blot membranes from mucin-containing samples after ultracentrifugation. (**a**) Post-staining with periodic acid-Schiff reagent (PAS). (**b**) Post-lectin staining of the same fractions. The fractions 3 and 4 (highlighted in red) are to be combined and kept for further analysis

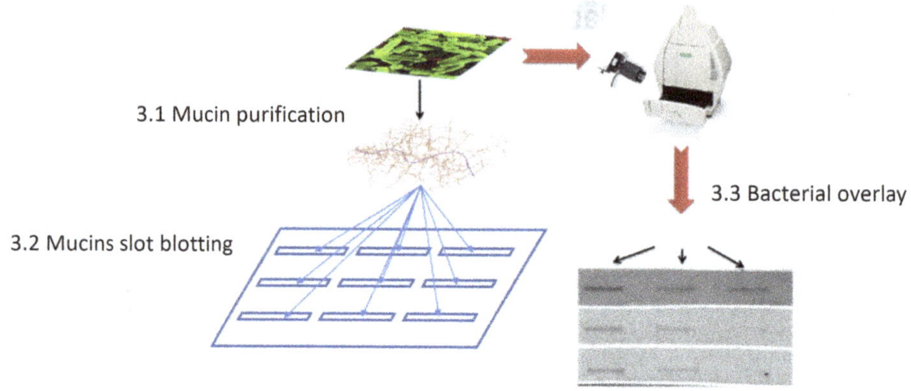

Fig. 2 Schematic view of the bacterial overlay

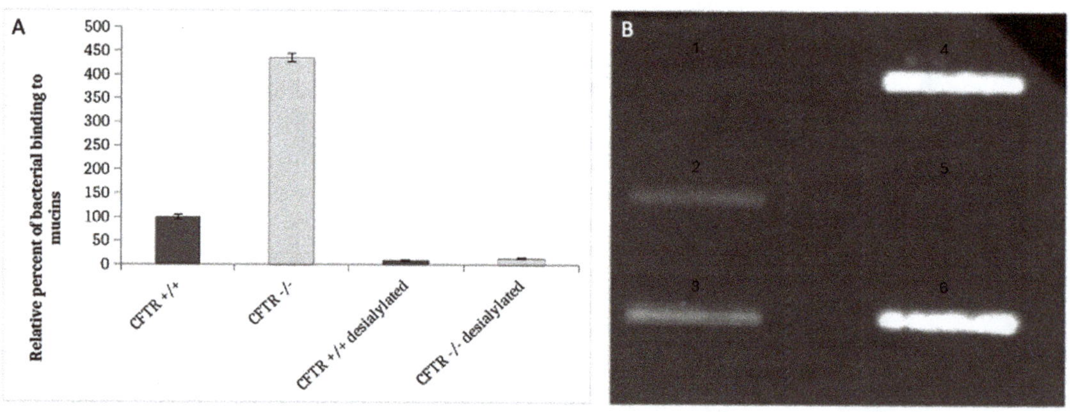

Fig. 3 (**a**) Quantification of bacterial binding (*Pseudomonas aeruginosa*) to chemically desialylated tracheal mucins or native mucins from CFTR +/+ or CFTR −/− piglets, indicating the role of sialic acids in this process. (**b**) The result obtained with samples of tracheal mucins from 2 CFTR +/+ piglets (1, 2 and 3) and 2 CFTR −/− piglets (4, 5 and 6)

assessed (Fig. 3). This method facilitates mucin manipulation and allows to study the implication of their glycans in bacterial adhesion.

2 Materials

All solutions should be prepared using ultrapure water (18 MΩ-cm at 25 °C) along with analytical-grade reagents. All reagents should be prepared and stored at room temperature unless specified otherwise. It is essential to adhere diligently to waste disposal regulations when discarding waste materials. Mucins can be purified from snap-frozen tissues or lyophilized tissues.

2.1 Mucin Purification

1. Tissue solubilizing buffers: 4 M guanidine chloride solution (*see* **Note 1**). Weigh 192 g of guanidium chloride (at least with a titration of 98%) in a volumetric flask of 500 mL (*see* **Note 2**). Add 200 mL of phosphate-buffered saline (PBS) 1X at pH 7.4. To prepare 1 L of PBS, weigh 8 g of NaCl, 0.2 g of KCl, 1.44 g of Na_2HPO_4, and 0.24 g of KH_2PO_4. Store at 4 °C.
2. Add extemporaneously cOmplete™ protease inhibitor cocktail with one tablet for up to 50 mL. Store at 4 °C.
3. Dithiothreitol (DTT).
4. Isopycnic density gradient centrifugation: cesium chloride for molecular biology (*see* **Note 3**).
5. Beckman Opti-Seal tubes used in the type 70.1 Ti rotor and a LE80 K ultracentrifuge (Beckman Coulter, Life Sciences).
6. 50 mL conical centrifuge tubes (Falcon™).
7. Dialysis tubing, cellulose membrane with a molecular weight cut-off of up to 20 K.

2.2 Slot Blot

1. Bio-Dot SF microfiltration of 48 wells (Bio-Rad).
2. A pumping system adapted to the Bio-Dot.
3. Nitrocellulose (0.2 μm pore size) or polyvinylidene fluoride (PVDF) membranes.
4. Solutions for periodic acid and Schiff reagent (PAS). Periodic acid solution: 1% (V/V) periodic acid in 3% of acetic acid (V/V) (*see* **Note 4**). Stop solution: 0.5% (W/V) sodium metabisulfite in 10 mM HCl (make it fresh). Schiffs' reagent (commercially available).
5. Lectins coupled to horseradish peroxidase.
6. Chemiluminescent detection reagents.
7. Tris-buffered saline (TBS 10×): 1.5 M NaCl, 0.1 M Tris–HCl, pH 7.4.
8. TBS containing 0.05% Tween-20 (TBST).
9. Blocking solution: 5% bovine serum albumin in TBS. Store at 4 °C.

10. Bioimager set to an excitation wavelength of 350 nm and an emission wavelength of 460 nm.

2.3 Bacterial Overlay

2.3.1 Bacterial Overlay to Determine Mucin Adhesion Capacity of Bacterial Strains or Species

1. Bio-Dot SF microfiltration of 48 wells (Bio-Rad).
2. A pumping system adapted to the Bio-Dot.
3. Nitrocellulose (0.2 μm pore size) membranes.
4. Phosphate-buffered saline (PBS) 1X at pH 7.4.
5. 4 M guanidinium chloride (optional).
6. Pierce™ Protein Free Blocking Buffer (PFBB).
7. 1.5 mL microtubes.
8. Microcentrifuge.
9. 4′,6-Diamidino-2-phenylindole dihydrochloride (DAPI).
10. Sealing film (Parafilm®).
11. Tween 20.
12. Bioimager set to an excitation wavelength of 350 nm and an emission wavelength of 460 nm.
13. Bacterial stock and their corresponding media (*see* **Note 5**).

2.3.2 Chemical Desialylation of Mucins

1. Trifluoroacetic acid.
2. 1.5 mL microtubes.
3. Microcentrifuge.
4. Water bath, heating block, or incubator set at 80 °C.

2.3.3 Enzymatic Desialylation of Mucins

1. 1.5 mL microtubes.
2. Microcentrifuge.
3. Neuraminidase (e.g., from New England Biolabs).
4. Water bath, heating block, or incubator set at 37 °C.

2.3.4 Epitope Masking by Lectins

1. Appropriate lectin.
2. 1.5 mL microtubes.
3. Microcentrifuge.
4. Neuraminidase.
5. Water bath, heating block, or incubator set at 37 °C.

3 Methods

Carry out all procedures at room temperature unless otherwise specified.

3.1 Mucin Solubilization and Ultracentrifugation

1. Solubilize the tissues in the tissues solubilizing buffer (12 mL) containing protease inhibitor cocktail in 50 mL conical centrifuge tube (Falcon™).
2. Use a cold mortar and pestle to thoroughly homogenize the tissues.
3. Add enough DTT to obtain a final concentration of 10 mM.
4. Incubate at least 24 h at around 20 °C.
5. Weigh the CsCl to reach a density of 1.40 g/mL and manually shake the tube until the CsCl is fully solubilized.
6. Transfer the samples to the ultracentrifuge tubes. Before centrifugation, ensure that the tubes are equilibrated by including the tube caps and adjusting the weight (with sample leftover or solubilizing buffer) if necessary.
7. Samples are placed in a pre-cooled rotor (previously cooled to 4 °C) and centrifuged at 4 °C for 72 h at 415,000 g.
8. After centrifugation, collect 500 μL fractions from the top of the tube. Weigh 50 μL of each fraction to monitor changes in density between fractions. Transfer 20 μL of each fraction to a nitrocellulose membrane using a slot blot followed by PAS staining to identify the mucin-rich fraction.
9. The mucin-containing fractions are pooled and transferred to dialysis tubing. The tubes are then immersed in water at 4 °C for at least 3 days, with the water changed twice a day.
10. Transfer the dialyzed samples to 50 mL conical centrifuge tubes (Falcon™), and freeze them before lyophilization. Lyophilization may take 1 or 2 days, depending on the model of the freeze dryer and the volume of samples. The powder obtained can be stored at room temperature until needed.

3.2 Slot Blot and Staining

1. Transfer 20 μL of each fraction to the nitrocellulose or PVDF membrane using the Bio-Dot SF microfiltration wells (*see* **Note 6**). When all the solution has been aspirated, open the Bio-Dot and transfer the membrane to water (*see* **Note 7**).

3.2.1 PAS Staining

1. For PAS staining, the membrane is incubated in the periodic acid solution for 30 min.
2. The membranes are washed three times for 2-min in distilled water and three times for 2-min washes in stop solution.
3. Incubate the membrane in Schiff's reagent for 15 min.
4. Perform three 2-min washes in the stop solution.
5. Before drying, consider photographing the membrane as the contrast is at its maximum at this stage.

3.2.2 Lectin Staining

1. Block the membranes with a blocking solution for 1 h.
2. Add appropriate lectin to the membranes and incubate for 2 h (*see* **Note 8**).
3. Wash three times with TBST, 5 min each time.
4. Rinse three times with deionized water before incubating with the chemiluminescent detection reagents as described by the manufacturer.

3.2.3 Pooling of Fractions

1. The density of mucins is usually comprised between 1.35 and 1.45 g/mL. Combine the fractions with such densities, after ensuring they are also the pinker fractions (Fig. 1a, fractions 3 and 4) or the more intense fractions after lectin detection (Fig. 2b) and transfer the resulting solution into dialysis tubing prepared according to the manufacturer's instructions (*see* **Note 9**).
2. Lyophilize samples and store them in a dry place at room temperature for future use.

3.3 Bacterial Overlay

The full process of this experiment is summarized in Fig. 2.

3.3.1 Bacterial Overlay to Determine Mucin Adhesion Capacity of Bacteria

1. Inoculate overnight bacterial cultures in 5 mL of fresh culture media at an OD600 of 0.1 and incubate the subcultures at the appropriate temperature with appropriate shaking (*see* **Note 10**) to reach an OD600 of 0.5.
2. Solubilize lyophilized purified mucins at a concentration of 1 μg/μL in PBS or in 4 M guanidine chloride solution (see Subheading 2.1, **item 1**).
3. Transfer 20 μL of each mucin to be tested to the nitrocellulose membrane using the Bio-Dot SF microfiltration wells. When all of the solution has been aspirated, open the Bio-Dot and transfer the membrane to PFBB for 1 h.
4. Centrifuge the full volume of bacterial subcultures at 3000 *g* in an Eppendorf tube for 2 min and discard the supernatant.
5. Wash bacteria by resuspending the bacterial pellet with 1 mL PBS and centrifuge at 3000 g for 2 min before discarding the supernatant. Repeat this step twice.
6. Resuspend bacteria with 2.5 μg/mL DAPI in PBS and incubate for 15 min in the dark at room temperature without shaking (*see* **Note 11**).
7. Remove excess DAPI by centrifugation at 3000 g for 2 min and discard the supernatant.
8. Wash bacteria by resuspending the bacterial pellet with 1 mL PBS and centrifuge at 3000 g for 2 min before discarding the supernatant. Repeat this step twice (*see* **Note 12**).

9. Suspend bacteria in 3 mL of PFBB and lay the obtained solution to the membrane. Incubate the membrane with labeled bacteria for 1 h in the dark at room temperature without shaking.
10. Wash the membrane three times in 15 mL of PBS containing 0.5% Tween 20 under gentle shaking at room temperature.
11. Detect bound fluorescent bacteria using a bioimager set to an excitation wavelength of 350 nm and an emission wavelength of 460 nm.

3.3.2 Bacterial Overlay to Determine Mucins-Bound O-Glycans Critical for Bacterial Binding

Chemical Desialylation of Mucins

1. Add 100 μL of ultrapure water to 0.1 mg of purified mucins to reach a concentration of 1 mg/mL in a microtube of 1.5 mL.
2. Mix thoroughly by vortexing for 1 min and spin down by short spin centrifugation (*see* **Note 12**).
3. Add 100 μL of 100 mM trifluoroacetic acid (TFA) in ultrapure water, mix briefly by vortexing, and spin down by short spin centrifugation.
4. Incubate the samples for 1 h at 80 °C in a water bath, a heating block, or an incubator.
5. Pursue as described in Subheading 3.3.1, **step 3**. Potential results are illustrated in Fig. 3.

Enzymatic Desialylation of Mucins

1. Add 180 μL of ultrapure water to 0.1 mg of purified mucins to reach a concentration of 1 mg/mL in an Eppendorf tube of 1.5 mL.
2. Mix thoroughly by vortexing for 1 min and spin down by short spin centrifugation (*see* **Note 13**).
3. Add 20 μL of NEB buffer 1, mix briefly by vortexing, and spin down by short spin centrifugation.
4. Add 2 μL of neuraminidase to reach a concentration of 500 U/mL (*see* **Note 14**).
5. Incubate the samples for 14 h at 37 °C in a water bath, a heating block or an incubator.
6. Pursue as described in Subheading 3.3.1, **step 3**.

Masking of Glycan Motives by Lectins

1. Perform bacterial overlay as described in Subheading 3.3.1 until Subheading 3.3.1, **step 3**.
2. Dilute the appropriate lectin (*see* **Note 15**) at a concentration of 20 μg/mL in 3 mL of PFBB and mix by gently pipetting up and down.
3. Remove PFBB from the membrane and cover the membrane with diluted lectin (*see* **Note 16**).

4. Wash the membrane three times in 15 mL of PBS containing 0.1% Tween 20 under slight shaking at room temperature.
5. Pursue with Subheading 3.3.1, **step 4.**

Bacterial Overlay to Determine whether Isolated or Pooled O-Glycans Are Critical for Bacterial Binding

1. Perform bacterial overlay as described in Subheading 3.3.1 until **step 8**.
2. Dilute the *O*-glycans to be tested at a concentration of 5 mg/mL in PFBB (*see* **Note 17**), mix by gently pipetting up and down.
3. Resuspend bacteria in the *O*-glycans containing PFBB solution and keep them for 1 h at room temperature in the dark.
4. Lay the obtained solution to the membrane. Incubate the membrane with *O*-glycan-coated labeled bacteria for 1 h in the dark at room temperature without shaking (*see* **Note 16**).
5. Pursue as described in Subheading 3.3.1, **step 10**.

4 Notes

1. Solubility in water is up to 6 M at room temperature or up to 8 M at 35 °C to form a stock solution. The solubilization process of the powder is endothermic, which means that the solution is initially cold as heat is absorbed during dissolution.
2. It is possible to use cheaper but less refined guanidinium chloride. To detect its lower quality, observe the solution after solubilization; if it appears cloudy, preliminary purification is necessary. Use 10 g of activated charcoal, either powdered or granular, per liter of solution. Incubate overnight with gentle stirring, first filter through a double layer of filter paper. Then pass the solution through a 0.22 μ nitrocellulose filter. After this treatment, the molarity of guanidinium chloride in water can be calculated from the density as follows: Molarity = $(\rho\text{-}1.003)/0.02359$. ρ is the density in grams per mL.
3. The amount of cesium chloride required to obtain the correct density can be calculated using the following equation: $m = V(1.347\rho\text{-}0.0318\ M\text{-}1.347)$, m is the CsCl needed in grams, V is the final volume, M is the molarity of the guanidinium chloride (4 M), and ρ is the density in grams per mL. The density, measured in grams per mL, should be assessed, as it may vary depending on the type of tissue, and adjusted if necessary.
4. Do not keep more than 1 month at 4 °C.
5. Bacterial adhesins are either membrane-bound or associated with surface appendages such as pili or flagella. To study their role in bacterial adhesion to mucins, a classical genetic approach is essential. This involves testing wild-type

(WT) strains alongside mutants lacking the adhesin and mutants complemented with a plasmid expressing the adhesin. To enhance the signaling response, mutants that overproduce adhesins would be beneficial, either through repressor mutations or the introduction of multicopy plasmids for adhesin expression.

6. PVDF membranes typically produce less background noise than nitrocellulose but need to be wetted in methanol for a few seconds before transfer to an aqueous-based buffer. Nitrocellulose membranes, on the other hand, require only a few seconds in an aqueous-based buffer before use.
7. Some fractions may be difficult to filter through the membrane and may require additional aspiration time. If possible, adjust the aspiration power accordingly. After 10 min of aspiration, remove the solution remaining in the wells by pipetting to prevent inadvertent spreading of the sample across the well.
8. *Helix pomatia* agglutinin (HPA), *Ulex europaeus* agglutinin I (UEA-I), and wheat germ agglutinin (WGA) each exhibit binding affinities for a wide range of mucins found in different tissue types. Specifically, HPA targets terminal α-*N*-acetylgalactosamine (α-GalNAc), UEA-I binds selectively to α-L-fucose residues, and WGA recognizes both *N*-acetylglucosamine and sialic acid residues.
9. Dry cellulose membranes used for dialysis must be boiled for 10 min in a solution containing 2% (w/v) sodium bicarbonate and 1 mM EDTA (pH 8) to remove glycerol. After boiling, rinse the membranes thoroughly with water. Once prepared, the membranes can be stored at 4 °C until needed. Cellulose acetate membranes are commercially available wet and only require thorough rinsing with water to remove the sodium azide used for preservation. The membranes must not be allowed to dry out after wetting, as drying can reduce the pore size.
10. Appropriate bacterial conditions should be adapted to the bacterial strain. To optimize specific adhesion to mucin-coated slots, bacterial suspensions ranging from 10^5 to 10^8 CFU/mL must be loaded.
11. Bacterial labeling can use fluorophores targeting DNA (e.g., DAPI, Syto9), membranes (e.g., PKH6), cell surface proteins (e.g., Alexa Fluor coupled ligands, antibodies), or fluorescent proteins such as GFP expressed by the bacteria.
12. Some bacterial species, such as *Pseudomonas aeruginosa*, do not pellet well after washing and labeling; if necessary, adjust the subculture volume during Subheading 3.3.1. to retain sufficient bacteria after these washing steps.

13. Depending on their source, mucins won't necessarily solubilize at this step, nevertheless, this does not affect the efficiency of the desialylation.
14. Other exo-glycosidases like ß-*N*-acetylglucosaminidase or α-fucosidase may be used to characterize the role of other glycans.
15. Masking of fucose epitopes can be performed by using *Ulex europaeus* agglutinin 1 (UEA1) for alpha1,2-linked fucose residues or *Aleuria aurantia* (AAL) for fucose α-1,6 *N*-acétylglucosamine motif or fucose α-1,3 galactose motif. Masking of sialic acid residues can be performed using *Maackia amurensis* lectin II (MALII) for α2–3 linked sialic acids and elderberry bark lectin (SNA) for α2–6 linked sialic acids, respectively.
16. To avoid liquid escaping the membrane, it might be easier to lay the membrane on sealing film (Parafilm ® M); leaving about 1 cm of film around the membrane.
17. Due to the high concentrations needed for this experiment, this will be rarely performed on isolated glycans; pooled glycans with different characteristics (for instance, neutral versus sialylated) will be favored. The volume of solution should be kept as small as possible as well as the surface of the membrane.

References

1. Martens EC, Neumann M, Desai MS (2018) Interactions of commensal and pathogenic microorganisms with the intestinal mucosal barrier. Nat Rev Microbiol 16:457–470
2. Linden SK, Sutton P, Karlsson NG et al (2008) Mucins in the mucosal barrier to infection. Mucosal Immunol 1:183–197
3. Derrien M, van Passel MWJ, van de Bovenkamp JHB et al (2010) Mucin-bacterial interactions in the human oral cavity and digestive tract. Gut Microbes 1:254–268
4. Rossez Y, Maes E, Lefebvre Darroman T et al (2012) Almost all human gastric mucin *O*-glycans harbor blood group A, B or H antigens and are potential binding sites for *Helicobacter pylori*. Glycobiology 22:1193–1206
5. Robbe C, Capon C, Coddeville B et al (2004) Structural diversity and specific distribution of O-glycans in normal human mucins along the intestinal tract. Biochem J 384:307–316
6. Luis AS, Hansson GC (2023) Intestinal mucus and their glycans: a habitat for thriving microbiota. Cell Host Microbe 31:1087–1100
7. Thornton DJ (2004) From mucins to mucus: toward a more coherent understanding of this essential barrier. Proc Am Thorac Soc 1:54–61
8. Davies JR, Carlstedt I (2000) Isolation of large gel-forming mucins. In: Glycoprotein methods and protocols. Humana Press, Totowa, pp 3–13
9. Fass D, Thornton DJ (2023) Mucin networks: dynamic structural assemblies controlling mucus function. Curr Opin Struct Biol 79: 102524
10. Ringot-Destrez B, D'Alessandro Z, Lacroix J-M et al (2018) A sensitive and rapid method to determin the adhesion capacity of probiotics and pathogenic microorganisms to human gastrointestinal mucins. Microorganisms 6:49

Chapter 16

Measuring and Analyzing Bacterial Movement in Mucus

Chris Viets and Corey A. Stevens

Abstract

Humans are colonized by trillions of microbes that compose the human microbiome. Much of the microbiome inhabits the mucus layers. Mucus layers, covering digestive, reproductive, ocular, and respiratory tracts, are viscous networks consisting mainly of water and mucin glycoproteins. Mucins assemble into a dense, cross-linked network that can affect bacterial swimming patterns, and studying this behavior provides valuable insights into how the body regulates interactions with both harmful and beneficial microbes. Here we present the use of time-lapse imaging to track individual bacterial cells within mucin and discuss techniques for accurately extracting cell trajectory data from these images. By integrating theoretical and experimental approaches, we also describe how to quantify bacterial movement in terms of speed, persistence, and randomness.

Key words Cell tracking, Mucin, Mucus, Fluorescence microscopy, Bacteria, Cell motility, Diffusion, Microbiome

1 Introduction

The human body is home to trillions of microbes that form the human microbiome, many of which reside in the mucus layers lining various tracts like the digestive, reproductive, ocular, and respiratory systems. Mucus, composed primarily of water and mucin glycoproteins, creates a dense, cross-linked network that can significantly impact the swimming patterns of bacteria. Understanding these interactions offers valuable insight into how the body manages its complex relationships with both harmful and beneficial microbes.

In this methods chapter, we detail the utilization of time-lapse imaging to track the movement of individual bacterial cells within mucin. We outline the techniques for accurately extracting cell trajectory data from these images, a crucial step in understanding bacterial behavior within this complex environment. Furthermore, by integrating theoretical and experimental approaches, we will elaborate on the methodologies to quantify bacterial movement

Inka Brockhausen (ed.), *Dynamics of Bacteria-Mucus Interactions*, Methods in Molecular Biology, vol. 2942,
https://doi.org/10.1007/978-1-0716-4627-4_16,

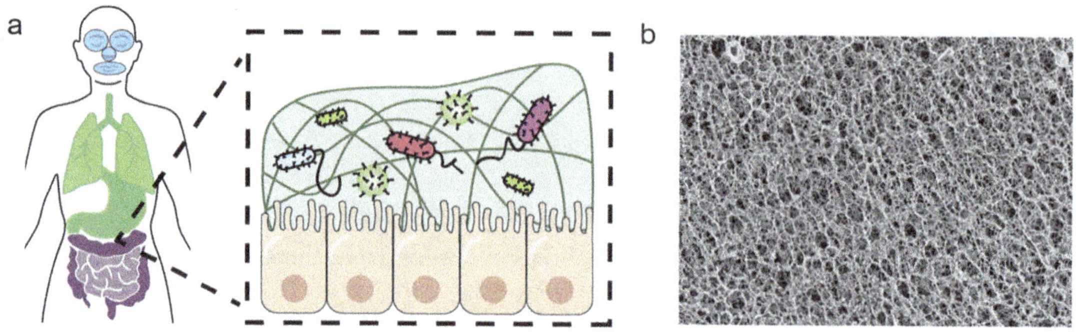

Fig. 1 (**a**) A schematic of where mucin is found in the human body. Mucin serves as an interface for epithelial cells to interact with microbiota in their environment. (**b**) A cryo-scanning electron microscopy (cryo-SEM) image of 1 wt% intestinal mucin (MUC2). Mucin polymers cross-link to form a mesh-like, hydrogel structure

parameters, including speed, persistence, and randomness. These analyses provide a framework for comprehensively characterizing bacterial motility in mucus and pave the way for future studies investigating the intricate relationship between the microbiome and its host environment.

Time-lapse imaging has led to the development of powerful computational methods for studying cell-substrate dynamics. Tracking algorithms that follow individual particles, organelles, and cells have become especially widespread in extracting motility data from time-lapse images, with applications in probing a substrate's microrheological properties, visualizing biochemical processes, and analyzing cell swimming behavior [1–4]. By providing trajectory measurements based on a series of time-lapse images, tracking algorithms yield kinematic results about the movement of a cell or particle through a substrate.

Cell tracking has served as an important tool in studying interactions between bacteria and mucin, the major biogel-forming, glycoprotein component of mucus (Fig. 1). By applying tracking algorithms to free-swimming bacteria and immotile beads in mucosal environments, previous studies have investigated the permeability of cervical mucus by bacteria [5] and have shown that mucin prevents the formation of biofilms by flagellated bacteria [6]. Bacterial cell tracking in mucus as outlined here can also be repurposed to larger cells such as lymphocytes, whose migration is reported to depend on the viscoelastic properties of the substrate [7, 8].

1.1 Quantitative Analysis of Bacterial Motility

Motility of *Escherichia coli* has been shown to follow a "run and tumble" pattern: cells move in straight-line, constant-velocity "runs" that are interspersed with brief "tumbles" in which the cell halts and randomly changes direction [9]. A tumble occurs when the bacterial flagella switch rotation directions, causing them to temporarily unbundle; this behavior allows the cell to respond chemotactically in the presence of a chemoattractant gradient by modulating its tumbling frequency [10]. Other bacterial species display similar types of motion: *Rhodobacter sphaeroides*, for example, swims in straight lines punctuated by stops during which the cell uses its own Brownian motion to randomly change direction [11]. In nearly all cases, bacterial cell motility includes an element of directional randomness, and the degree of randomness can be quantified by extracting the relevant parameters from cell track data.

Cell tracking provides measurements of a cell's x, y, and (if available) z positions as a function of time, $\vec{r}(t)$. The mean squared displacement (MSD) serves as a measurement for the spatial coverage of tracked cells. Assuming the ith cell is on-screen from time frames $t_0^{(i)}$ to $t_f^{(i)}$, the cell's time-averaged MSD is

$$\overline{r_i^2(\tau)} = \frac{1}{t_f^{(i)} - t_0^{(i)} - \tau + 1} \sum_{t=t_0^{(i)}+\tau}^{t_f^{(i)}} \left|\vec{r_i}(t) - \vec{r_i}(t-\tau)\right|^2 \quad (1)$$

Here, t indexes the frames in the video and τ refers to a time difference. The final MSD is computed by taking the ensemble average:

$$\langle r^2(\tau)\rangle = \frac{1}{N}\sum_{j=1}^{N}\overline{r_j^2(\tau)} \quad (2)$$

In general, the MSD obeys the power law

$$\langle r^2(\tau)\rangle = 2nMt^{\alpha} \quad (3)$$

where n is the spatial dimension (e.g., $n = 2$ for 2D microscopy images) and M is the cell motility coefficient [12]. The α parameter measures a cell's ability to travel from its starting point compared to a random walk, where cells following a random walk have $\alpha = 1$ [13]. The extreme case of $\alpha = 2$ occurs when cells travel in straight lines. The MSD is often plotted on a log-log scale so that it appears as a line with slope α.

For bacteria with run-and-tumble motion, a typical MSD profile has $\alpha \approx 2$ at short timescales (i.e., times comparable to the duration of a run), followed by a tapering to $\alpha \approx 1$ at longer timescales (Fig. 2). Such an MSD profile indicates that cells are engaging in a discrete random walk, taking straight-line steps that each have a random direction. In general, increased cell speed can contribute to a larger motility coefficient M, while greater directional persistence increases the α parameter, as directionally persistent cells can achieve large displacements over long times even when cell speed is low. The MSD therefore captures both short-term and long-term motility behavior and contains information about cell speed and directionality.

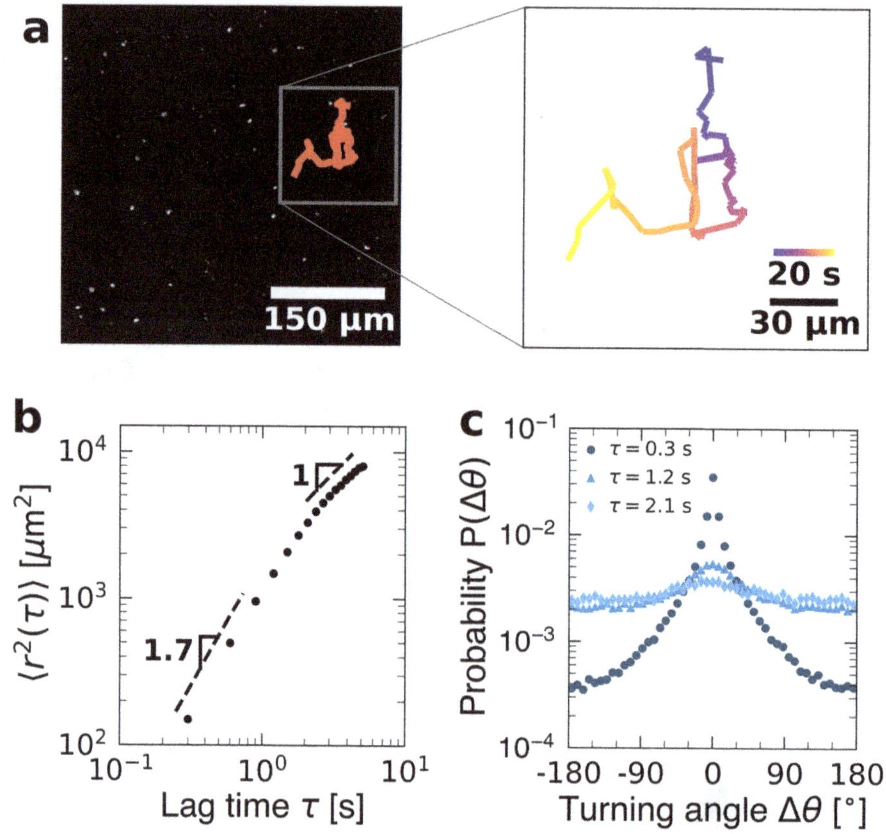

Fig. 2 Time-lapse images of GFP-expressing *E. coli* in 0.5 wt% mucin (MUC2) were taken at a rate of 200 frames per minute for 3 min. (**a**) A typical cell's trajectory taken from the time-lapse series. (**b**) MSD profile of the tracked cells. Cells exhibited directed motion at timescales shorter than 1 s, and random motion across longer timescales. (**c**) Turning angle distribution is peaked at 0° at a lag time of 0.3 s but becomes closer to uniform at longer lag times

To obtain a more complete picture of cell motility, cell speed and persistence should be measured directly. Like the MSD, mean cell speed $\langle v \rangle$ is computed as the population average of the time average:

$$\overline{v}_i = \frac{1}{t_f^{(i)} - t_0^{(i)} + 1} \sum_{t=t_0^{(i)}}^{t_f^{(i)}} \left| \vec{v}_i(\mathrm{t}) \right| \tag{4}$$

$$\langle v \rangle = \frac{1}{N} \sum_{j=1}^{N} \overline{v}_j \tag{5}$$

where $\vec{v}_i(t)$ is an n-component vector denoting the ith cell's velocity. Cell persistence is quantified by measuring the distribution of turning angles $\Delta\theta$ across a fixed span of time τ,

$$\Delta\theta = \arg\left(\vec{v}_i(t)\right) - \arg\left(\vec{v}_i(t-\tau)\right) \tag{6}$$

for all cells i and all times t. For cells with run-and-tumble motion, the turning angle distribution has a narrow peak at 0° at short timescales when cells engage in straight-line runs but becomes more uniform at longer timescales as each cell's directionality becomes random (Fig. 2c). In general, a narrower peak at 0° is indicative of increased persistence, as cells are more likely to maintain the same direction.

1.2 Correcting for Artifacts of Automated Cell Tracking

Automated cell tracking packages typically apply a Laplacian of Gaussian (LoG) filter to determine a cell's edges [14]. This method provides high-precision position measurements when all cells are sufficiently bright and far apart. However, when a cell spot briefly becomes too dim (e.g., by temporarily shifting out of the focal plane), the cell tracking software may fail to assign a track to the cell at those time points, leaving a gap in the trajectory called a track splitting artifact. Additionally, when the distance between two cells becomes comparable to the cell spot radius, the automated tracking software may swap their trajectories, giving rise to a track switching artifact. Typically, track switching artifacts can be prevented by prematurely ending tracks if they come within a predefined distance of each other. However, this approach gives rise to additional track splitting artifacts and prevents analysis of cell motility over longer timescales.

We provide a track analysis pipeline that corrects both artifacts without making this tradeoff by implementing track splitting and track linking algorithms. The track splitting algorithm splits a track if it belongs to two different cells: it searches for the presence of "breakpoints" in each track, locations where the track abruptly deviates from its past behavior. Conversely, the track linking algorithm stitches two tracks together if they belong to the same cell and occurs by a three-step process: (1) for each track, identify candidate tracks that could plausibly link to it; (2) select the best candidate track; and (3) interpolate the space between the two tracks. Running the splitting algorithm followed by the linking algorithm simultaneously corrects for both types of artifacts, first removing erroneous track switches and then mending tracks together in the correct configuration (Fig. 3).

2 Materials

2.1 Flow Cell Preparation

1. Glass microscope slides.
2. Glass coverslips.
3. Double-sided tape.
4. Valap (1:1:1 mixture of Vaseline, lanolin, and paraffin).

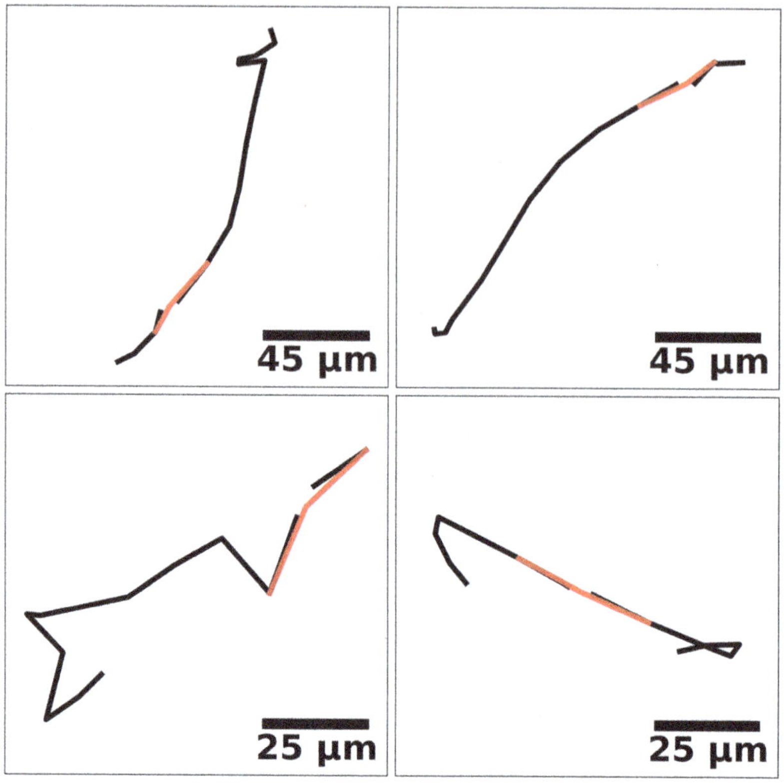

Fig. 3 The linking algorithm connects two tracks if they both belong to the same cell. The linking algorithm fits second-degree polynomials to the x and y position data to interpolate the links (shown in red) between the two tracks (shown in black), turning them into one track

2.2 Bacterial Imaging

1. Luria-Bertani (LB) broth.
2. Fluorescent bacteria.
3. Natively purified mucins (methods as previously described).

2.3 Track Data Analysis

A pipeline for cell trajectory postprocessing and analysis is provided at (https://github.com/cviets/cell_tracking). Subheading 3.6 contains a quick glossary of its functions. Users will need to install Python 3 and NumPy [15] if not already installed. Note that the pipeline can currently only accommodate two-dimensional cell tracking.

3 Methods

3.1 Flow Cell Construction

A diagram of the flow cell construction is shown in Fig. 4.

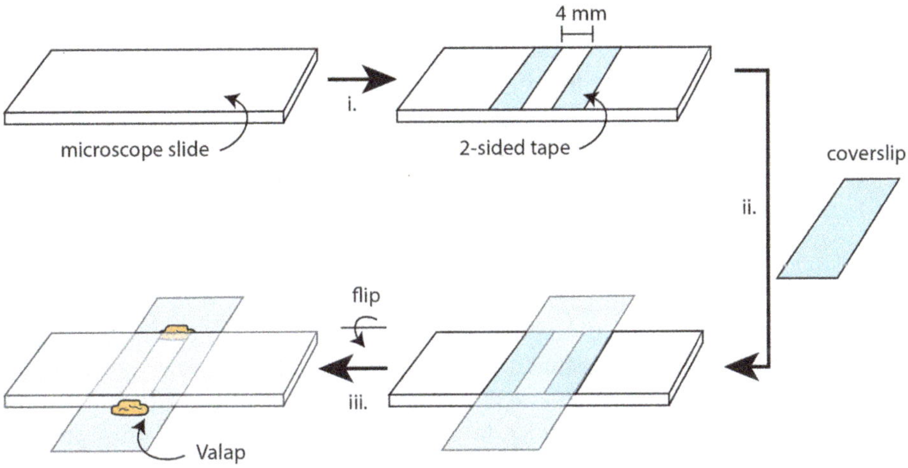

Fig. 4 A diagram of the flow cell used to image bacteria. (i) Place two strips of double-sided tape four millimeters apart on microscope slide. (ii) Place coverslip on tape and press firmly to stick. (iii) Fill with approximately 20 μL of solution and seal the open junctions with valap

1. Lay a glass microscope slide horizontally on a clean surface. Place two strips of double-sided tape 4 millimeters apart from each other, running vertically over the width of the glass slide.
2. Gently place a glass coverslip vertically on top of the tape. After this step, there should be a thin gap of air sandwiched between the glass slide and coverslip.
3. To seal the flow cell after adding bacteria, heat valap to 50 °C until liquid, and brush or drip onto junction. Valap should harden and seal immediately and is nontoxic to cells (*see* **Note 1**).

3.2 Preparing Bacteria for Imaging

1. Pick a single colony of bacteria and inoculate into 3 mL of LB broth in a 15 mL conical tube.
2. Incubate the liquid bacterial culture at 37 °C with shaking (220 rpm) overnight.
3. Dilute the overnight culture 1:100 into fresh LB and incubate at 37 °C with shaking (220 rpm) for 3 h.
4. After the 3-h incubation, dilute the bacteria 1:100 in LB.

3.3 Time-Lapse Imaging of Fluorescently Labeled Bacteria

1. Fill a flow cell with bacteria suspended in LB (*see* **Notes 2** and **3**).
2. Placing the flow cell under the microscope, estimate the number of bacteria on screen to determine whether bacteria concentration should be diluted. We recommend a cell density of around 40 bacteria on screen at a time. If necessary, dilute the cells in LB. Repeat **steps 1** and **2** until cell density is satisfactory, keeping in mind that **step 3** involves diluting the cell density with mucin and water.

3. Add water and mucin to the bacteria in LB so that the mucin concentration and bacterial cell density are as desired, and so that the LB is at 1× concentration.
4. Fill a new flow cell with the bacteria in mucin, and seal the edges with valap (*see* **Note 1**).
5. Take time series images for 3–5 min at a frame rate of 200 frames per minute or faster (*see* **Notes 4–6**).

3.4 Generating and Processing Track Data

Generating track data involves assigning time and position coordinates to all cells in a microscopy video. We perform this step in FIJI [16] using the TrackMate plugin [17]. Track data is then processed and analyzed using the provided Python analysis pipeline. The pipeline includes functions to de-drift tracks and correct for artifacts of automated tracking software.

1. Open microscopy images in FIJI and adjust brightness and contrast so that the cells appear as bright spots against a dark background. Perform automated cell tracking using TrackMate, and export cell track data in CSV format (*see* **Notes 7–9**).
2. If the cells are observed to drift, track de-drifting is strongly recommended. The most common method to de-drift track data is to measure the ensemble mean displacement at each time step and then subtract this value from each track's displacement. Hence, for an arbitrary track $\vec{r}^{(k)}(t)$,

$$\vec{r}^{(k)}_{\text{corrected}}(t) = \vec{r}^{(k)}(t) - \left\langle \vec{r}^{(i)}(t) \right\rangle_i \tag{7}$$

   where $\vec{r}^{(k)}(t)$ denotes the kth cell's trajectory.
3. Check for cell track splitting and switching artifacts by comparing a sample of tracks to the original time-lapse series in FIJI, visually inspecting whether each track follows exactly one bacterial cell. We recommend looking at the longest-duration tracks, which are more likely to contain tracking artifacts. Run track splitting and linking algorithms to correct for any observed artifacts (*see* **Notes 10** and **11**).

3.5 Analyzing Track Data

1. Compute the MSD as shown in Eq. (2) for all available lag times, averaging over the ensemble of cells at all time points. Fit the MSD to the power law in Eq. (3) to determine the motility coefficient M and the α parameter (*see* **Note 12**).
2. Measure the mean cell speed by taking the time derivative of cell positions and averaging over time points and the cell population, as in Eq. (5).
3. Compute the turning angle distribution as shown in Eq. (6). It is often desirable to compute the turning angle distribution for

two or more lag times to gain information about the short- and long-term behavior of the cell.

4. Altogether, M, α, mean cell speed, and the turning angle distribution provide information about cellular motility, especially when cells deviate from random walk behavior. For example, if M is small but $\alpha > 1$, cells may be swimming slowly but persistently, maintaining the same direction for long stretches of time, which can be confirmed using the mean cell speed and the turning angle distribution.

3.6 Track Analysis Software and Glossary of Functions

1. *Tracks(filename, x_column = 'E', y_column = 'F', time_column = 'I', id_column = 'C')*: Create a new *Tracks* object, which stores all relevant cell trajectory data in the input CSV file. The arguments after *filename* specify the columns in the CSV file containing position, time, and track ID data, with defaults set according to TrackMate's CSV outputs (*see* **Note 13**).
2. *Tracks.dedrift()*: De-drift all trajectories. Returns the displacement due to drift at each time point.
3. *Tracks.correct_artifacts(max_time_gap, max_distance)*: Split tracks that belong to two or more different cells and link tracks that belong to the same cell. Only tracks whose endpoints occur within *max_time_gap* and *max_distance* from each other will be considered for linking (*see* **Notes 10** and **11**).
4. *Tracks.turning_angles(dt)*: Get the turning angle distribution for a lag time of *dt.*
5. *Tracks.mean_speed()*: Get the ensemble mean speed of the tracked cells.
6. *MSD_analysis.MSD(Tracks.tracks):* Compute the mean squared displacement of the tracked cells as a function of time lag.

4 Notes

1. Failing to seal the flow cell with valap may result in sample evaporation and increased drift when imaging. When counting cell density on the microscope prior to taking images, however, valap is not needed.
2. It is possible to stack multiple layers of double-sided tape on top of each other prior to placing the glass coverslip. This setup may be desirable if imaging multiple *z*-stacks, for example. However, increasing the height of the flow cell may cause additional drift, especially near the edges of the flow cell.
3. Around 20 uL should be needed to fill the flow cell chamber.

4. For 2D imaging, take caution not to image the bacteria at a *z*-stack near the top or bottom of the flow cell. Bacteria near the glass surface tend to swim in circles, as the rotation of the flagella creates a centrifugal force when in close proximity with the liquid–solid interface [18].
5. Imaging for longer than 5 min is not recommended, as bacteria may begin gravitationally settling to the bottom of the flow cell over this duration. Moreover, bacteria tend to remain on screen for less than a minute at a time, so imaging for longer than 5 min is unlikely to provide additional data about long-term motility behavior.
6. All imaging was performed using a Zeiss Axiovert inverted fluorescent microscope at 20× magnification and recorded on a Hamamatsu Orca Flash 4.0 V3. However, any fluorescent microscope should be sufficient as long as the frame rate and magnification of imaging are known.
7. After adjusting brightness and contrast, FIJI's "Remove Background" function can be called to further darken the background, yielding better tracking results.
8. In some cases, image noise may produce spots of similar size to cells. FIJI's "Despeckle" function can be called to smooth over speckles caused by noise while preserving cell spots.
9. It is recommended to save the TrackMate XML file until data postprocessing and analysis are complete. The XML file allows users to reopen the TrackMate output in the TrackMate GUI. For instance, the TrackMate output may be needed when searching for cell tracking artifacts.
10. The *correct_artifacts* function prints a list of track pairs it will link before modifying any tracks. Users should verify the accuracy of these pairings in the TrackMate GUI before proceeding with artifact correction.
11. When bacterial cell density is high and bacteria quickly appear and disappear from the field of view, it may be difficult to visually identify cell tracking artifacts. In this case, we recommend that the parameters *max_time_gap* and *max_distance* be kept low. Making either of these parameters too large is likely to create links between tracks that do not belong together, yielding erroneous results about long-term bacterial motility.
12. The most straightforward method of fitting mean squared displacement data to the power law in Eq. (3) is by plotting the MSD on a log-log scale and fitting the data points to a line.
13. The *time_column* argument should specify the column that measures time in frames rather than seconds. Integer time measurements are needed to prevent rounding errors that may lead to inconsistencies in data analysis.

References

1. Vivante A, Brozgol E, Bronshtein I, Garini Y (2017) Genome organization in the nucleus: from dynamic measurements to a functional model. Methods 123:128–137
2. Fournier M, Leclerc P, Leray A et al (2023) Combined SPT and FCS methods reveal a mechanism of RNAP II oversampling in cell nuclei. Sci Rep 13(1):14633
3. Miné-Hattab J, Recamier V, Izeddin I, Rothstein R, Darzacq X (2017) Multi-scale tracking reveals scale-dependent chromatin dynamics after DNA damage. MBoC 28(23): 3323–3332
4. Alirezaeizanjani Z, Großmann R, Pfeifer V, Hintsche M, Beta C (2020) Chemotaxis strategies of bacteria with multiple run modes. Sci Adv 6(22):eaaz6153
5. Smith-Dupont KB, Wagner CE, Witten J et al (2017) Probing the potential of mucus permeability to signify preterm birth risk. Sci Rep 7(1):10302
6. Caldara M, Friedlander RS, Kavanaugh NL, Aizenberg J, Foster KR, Ribbeck K (2012) Mucin biopolymers prevent bacterial aggregation by retaining cells in the free-swimming state. Curr Biol 22(24):2325–2330
7. Oakes PW, Patel DC, Morin NA et al (2009) Neutrophil morphology and migration are affected by substrate elasticity. Blood 114(7): 1387–1395
8. Pittman M, Iu E, Li K et al (2022) Membrane ruffling is a mechanosensor of extracellular fluid viscosity. Nat Phys 18(9):1112–1121
9. Berg HC, Brown DA (1972) Chemotaxis in Escherichia coli analysed by three-dimensional tracking. Nature 239(5374):500–504
10. Turner L, Ryu WS, Berg HC (2000) Real-time imaging of fluorescent Flagellar filaments. J Bacteriol 182(10):2793–2801
11. Wadhwa N, Berg HC (2022) Bacterial motility: machinery and mechanisms. Nat Rev Microbiol 20(3):161–173
12. Berg HC (1993) Random walks in biology. Princeton University Press, p 168
13. Metzler R, Jeon JH, Cherstvy AG, Barkai E (2014) Anomalous diffusion models and their properties: non-stationarity, non-ergodicity, and ageing at the centenary of single particle tracking. Phys Chem Chem Phys 16(44): 24128–24164
14. Ershov D, Phan MS, Pylvänäinen JW et al (2022) TrackMate 7: integrating state-of-the-art segmentation algorithms into tracking pipelines. Nat Methods 19(7):829–832
15. Harris CR, Millman KJ, van der Walt SJ et al (2020) Array programming with NumPy. Nature 585(7825):357–362
16. Schindelin J, Arganda-Carreras I, Frise E et al (2012) Fiji: an open-source platform for biological-image analysis. Nat Methods 9(7): 676–682
17. Tinevez JY, Perry N, Schindelin J et al (2017) TrackMate: an open and extensible platform for single-particle tracking. Methods 115:80–90
18. Lauga E, DiLuzio WR, Whitesides GM, Stone HA (2006) Swimming in circles: motion of bacteria near solid boundaries. Biophys J 90(2):400–412

Chapter 17

Adhesion of *Pseudomonas aeruginosa* to Human Outer Layer Airway Mucins

Burkhard Tümmler

Abstract

The adhesion of *Pseudomonas aeruginosa* to submaxillary and airway mucins is subject to large intra- and interclonal bacterial diversity. Thus, a mucin adhesion study should include numerous bacterial strains that reflect the population biology of the taxon and/or the habitat of interest. Human airway mucins are isolated from induced sputa collected from healthy individuals or people with chronic lung diseases such as bronchiectasis, cystic fibrosis, or chronic obstructive pulmonary disease. In this protocol, bacteria are exposed to mucin-coated microplates. Adhesion is quantified by colony forming units.

Key words Adhesion, Airways, Bronchiectasis, COPD, Cystic fibrosis, Mucin, *Pseudomonas aeruginosa*, Respiratory tract

1 Introduction

Surface epithelia throughout the body of vertebrates are covered by mucus, a protective secretion that serves as a physical barrier between the epithelial cell plasma membrane and the extracellular environment [1]. Mucus layers vary widely in composition and structure; for example, they are thick and adherent to the epithelium in the gut, but thin and mobile in the airway [2]. Macroscopically, mucus is a non-Newtonian, thixotropic gel; under low shear, mucus behaves like an elastic solid, and under high shear, mucus behaves like a viscous liquid and eventually deforms irreversibly [3, 4]. Mucus consists of 97% water and 3% solid [2] (mucins, DNA, lipids, ions, proteins, cells, and cellular debris), with mucins being the major solid constituent.

The gel-forming mucins are characterized by a high frequency of the hydroxy-amino acids serine and threonine, along with proline (PTS domain), and are glycosylated predominantly by the *O*-glycans, which typically make up more than 80% of the mass of a mucin [5]. These central O-glycosylated mucin repeat domains are

Inka Brockhausen (ed.), *Dynamics of Bacteria-Mucus Interactions*, Methods in Molecular Biology, vol. 2942, https://doi.org/10.1007/978-1-0716-4627-4_17,

often interrupted by 100-amino-acid-long non-glycosylated CysD knot domains [5]. Humans mainly express four mucin proteins, namely, MUC2 and MUC6 in the gastrointestinal tract; MUC5AC and MUC5B in saliva and the respiratory tract [6]. MUC5AC and MUC5B form linear N-terminal covalent dimers whereby MUC5AC dimers moreover non-covalently aggregate into net-like structures [5].

Human airways are continuously exposed to microbes, but in the healthy state, the alveolar compartment is sterile and the conducting airways are harboring only a tiny fraction of the commensal oropharyngeal flora [7, 8]. Thanks to the mucociliary escalator bundles of MUC5B polymers clean the tracheobronchial surface and remove bacteria [9–12]. Conversely, upon infection or in chronic lung diseases such as bronchiectasis, cystic fibrosis (CF), or chronic obstructive pulmonary disease (COPD), MUC5AC becomes more abundant and together with MUC5B can organize a stratified mucus layer that can separate bacteria from the epithelial surface [5, 6, 13, 14].

The global role of mucins to control bacterial infection has been demonstrated, but the binding epitopes that mediate the interaction between mucin and bacterium have only been resolved in a few cases (e.g., [15, 16]). Hereunder we describe microplate assays to assess the adhesion of the opportunistic pathogen *Pseudomonas aeruginosa* to human respiratory gel-forming mucin. Being a ubiquitous aquatic microorganism, the nosocomial *P. aeruginosa* may cause life-threatening airway infections in ventilated patients in intensive care units or may sustain chronic lung infections in people with COPD, bronchiectasis, or CF [17]. This assay was designed to quantify bacterial binding to mucin and to resolve the carbohydrate epitopes that are recognized by the bacterial strain. The recognized mucin epitopes are subject to substantial inter- and intraclonal bacterial diversity. Thus, in contrast to the common approach to focus on just one or a few reference strains such as PAO1 or PA14 in case of *P. aeruginosa*, a study should include a sufficient number of strains that reflect the population biology of the taxon and/or the habitat of interest [17]. Knowledge of the bacterial genome sequence is desirable [18], but at least the multilocus sequence type of the bacterial isolate should be known [19].

Most commercially available mucins do not form a hydrogel and thus cannot reproduce the properties of native mucus, which limits the usefulness of commercial mucins for experiments mimicking the barrier properties of native mucus towards pathogens [20–22]. Hence, the author's laboratory decided to isolate human airway mucin directly from induced sputum and to follow the protocol by Snyder et al. from the 1980s that yields a single high-molecular-weight peak in analytical ultracentrifugation and shows no low-molecular-weight protein ($< 10^5$) in the Coomassie stain of SDS-PAGE gels [23]. Proteins non-covalently bound to airway mucin by ionic interaction are removed by reversible modification of lysine residues with citraconic anhydride so that the *O*-glycosylated mucin can be precipitated with cetyltrimethylammonium bromide. The *O*-glycans should make up more than 70% of the mass.

The relative contents of MUC5B and MUC5AC in the purified mucin will depend on the health status of the sputum donor. Consequently, some general and disease-specific information about the sputum donor should be recorded (see below). If technically feasible at the site, the contents of MUC5B and MUC5AC may be quantified by stable isotope labeling mass spectrometry [24].

Binding assay (*see* **Note 1**): To assess bacterial adhesion to human airway mucin, viable bacteria of defined growth phase are exposed to a mucin-coated microplate. Bacterial binding is quantified by colony forming units (CFU). To discern the carbohydrate epitopes, soluble mono- or oligosaccharides compete with the mucin for bacterial binding. Commercially available mucins such as the bovine submaxillary gland mucin should be included as an internal standard.

The binding assay will not reveal any classical affinity constants of binary or multiple equilibria but rather will provide semiquantitative information about the bacterial capacity to adhere to the *O*-glycans of mucins. Topics of microbiological interest could be the following:

- The search for bacterial mucin receptors by screening libraries of isogenic bacterial mutants.
- The impact of habitat, nutrients, and growth phase on mucin adhesion.
- Intra- or interclonal comparisons of lineages to bind to airway mucin.
- Microevolution of mucin adhesion in serial clonal isolates from the same habitat.

2 Subjects and Materials

2.1 Human Subjects

1. Obtain approval to collect induced sputum from your local ethics committee.
2. Obtain written informed consent from each volunteer prior to collection of induced sputum.
3. Record: date and time of induced sputum collection; age, gender, and geographic origin of donor.

   Record in addition:

4. For people with COPD: diagnosis, disease status, and etiotype according to GOLD 2023 [25, 26].
5. For people with bronchiectasis: etiology, bronchiectasis severity index [27, 28].
6. For people with CF: *CFTR* genotype, treatment with CFTR modulators (yes/no), FEV1 CF percentile [29, 30] (FEV1, forced expiratory volume in 1 s).

2.2 Induced Sputum

1. Bronchodilator: salbutamol (100 μg/dose), with anti-static valved holding chamber.
2. Hypertonic saline (HTS) from pharmacy: 3% for expectorating participants, naïve to higher concentrations (if they are or have previously been prescribed 7%, giving them 3% may not work). The % HTS up to 7% may be decided on a case-by-case basis.
3. Nebulizer equipment (e.g., PARI BOY Classic).
4. Supplemental oxygen.
5. Pulse oximeter.
6. Peak flow meter or spirometer.
7. Small plastic containers for saliva and sputum.
8. Styrofoam container with dry ice.

2.3 Isolation of Human Airway Mucin

1. Bidistilled water.
2. Sterile isotonic saline (0.9% (w/v) NaCl).
3. 0.1 M NaCl.
4. 0.1 M EDTA.
5. 0.5 M Tris-HCl, pH 8.0.
6. 1 M NaOH.
7. 1% (w/v) sodium azide.
8. 20% (w/v) N-cetyl-N,N,N trimethyl ammonium bromide.
9. Citraconic anhydride.
10. Acetic acid.
11. Dialysis tubes (Visking cellulose type 20/32).
12. Centrifuge and tubes.
13. Magnetic stirrer.
14. pH meter.
15. Dry freezer.

All glassware, buffers, and saline solutions should be autoclaved prior to use.

2.4 Mucins, Mono-, and Oligosaccharides

1. Commercially available monosaccharides (*see* **Note 2**): fucose, galactose, mannose, N-acetylgalactosamine, N-acetylglucosamine, and neuraminic acid.
2. Oligosaccharides present in *O*-glycans and oligosaccharides present in *N*-glycans may be included for comparison (*see* **Note 3**).
3. Mucins (*see* **Note 4**).
   (a) Ovine submaxillary mucin (OSM) [31].
   (b) Asialo ovine submaxillary mucin (AOSM).
   (c) Porcine submaxillary mucin (PSM) [32].
   (d) Asialo porcine submaxillary mucin (APSM).
   (e) Bovine submaxillary mucin (BSM) [33].

2.5 Bacteria and Media

1. *P. aeruginosa* strains stored at −80 °C in 0.3 mL aliquots in micro-tubes at −80 °C in lysogenic broth (LB) supplemented in 17% (v/v) glycerol (*see* **Note 5**, [34–37]).
2. Lysogenic broth (LB): 15 g/L casein hydrolysate, 5 g/L yeast extract, 10 g/L NaCl, adjust to pH 7.0. Autoclave at 121 °C for 20 min, and store at 4 °C.
3. LB-agar plates: to generate solid medium, add 10 g/L agar. Autoclave at 121 °C for 20 min, and store at 4 °C.
4. M9 minimal medium:
   (a) 5×M9 stock solution. Per liter: 42.5 g $Na_2HPO_4 \times 2\ H_2O$, 15 g KH_2PO_4, 5 g NH_4Cl, 2.5 g NaCl. Adjust to pH 7.0 using 4 M NaOH. Autoclave at 121 °C for 20 min, and store at room temperature.
   (b) 100× trace elements stock solution. Per Liter: 100 mg $MnCl_2 \times 4\ H_2O$, 170 mg $ZnCl_2$, 43 mg $CuCl_2 \times 2\ H_2O$, 60 mg $CoCl_2 \times 6\ H_2O$, and 60 mg $Na_2MoO_4 \times 2\ H_2O$. Filter sterilize the solution and store in the dark at room temperature.
   (c) Iron chloride solution (50 mM in 100 mM citric acid, 100 mL): 1.35 g $FeCl_3 \times 6\ H_2O$ and 2.10 g citric acid. Filter sterilize and store in the dark at 4 °C.
   (d) Calcium chloride solution (100 mM, 100 mL): 1.47 g $CaCl_2 \times 2\ H_2O$. Autoclave and store at room temperature.
   (e) Magnesium sulfate solution (1 M, 100 mL): 24.6 g $MgSO_4 \times 7\ H_2O$. Autoclave and store at room temperature.
   (f) Carbon source: glycerol (10% (v/v), 100 mL): 10 mL. Freshly prepared, filter sterilize just before you set up the M9 medium (g).

(g) For 1 l of M9 medium, add in the following order under stirring (10–20 rpm): 200 mL M9 5× stock, 1 mL 100 mM $CaCl_2$, 10 mL trace salts 100× stock, 1 mL 1 M $MgSO_4$ (slowly add dropwise to avoid any precipitation), 1 mL iron chloride solution, and 100 mL glycerol (10%), and add bidistilled H_2O to a final volume of 1 L.

5. Phosphate-buffered saline (PBS). To determine 'colony forming units' (CFU), samples are diluted in steps of tenfold with PBS (8.0 g/L NaCl, 0.2 g/L KCl, 0.132 g/L $CaCl_2 \times 2\ H_2O$, 0.1 g $MgCl_2 \times 6\ H_2O$, 1.15 g Na_2HPO_4, and 0.2 g KH_2PO_4).

2.6 Mucin Binding Test

1. 96-well polystyrene U-bottom microplates.
2. Solution of 10 μg mucin/ml PBS (airway mucin or submaxillary mucin; see Subheading 2.4, **item 3**).
3. Glass Drigalski spatulas.
4. 1% (v/v) Triton × 100 in PBS.

3 Methods

3.1 Collection of Induced Sputum

1. Information sheet should be provided before the procedure (*see* **Note 6**).
2. Advise the participant to avoid eating 1 h prior to the procedure.
3. Advise participant to follow their normal medication schedule using their prescribed bronchodilator, if not prescribed salbutamol.
4. Have a styrofoam container with dry ice ready for storage of the collected sputum. Deposit the small plastic sputum container in dry ice.
5. Instruct the participant (if <18 years, the participant & caregiver) on how the sputum induction procedure will be performed including advice regarding any adverse symptoms they may experience. Hypertonic saline (HTS) will produce a cough in most participants. It may also taste unpleasant and cause some throat irritation. It is important that the administrator of the test dose of HTS explains these effects.
6. If required, the procedure can be paused (where the participant is tiring or needs to clear sputum), and a drink of water should be made available for the participant, as regular sips of water may help to make the taste more tolerable.
7. Measure baseline vital signs: oxygen saturation (SpO_2), respiratory rate, heart rate, temperature, and blood pressure. If diabetic, check when the participant's last meal was taken and a recent blood glucose result (<1 h). If the participant is

prescribed O_2, ensure that this is available and regularly check SpO_2 during the procedure. If oxygen saturation is <95%, or as deemed necessary by the medical personnel, participants will have continuous pulse oximetry throughout the procedure. If oxygen saturation is <90%, *induced sputum must not be collected. Stop the procedure.*

8. Measure pre-sputum induction baseline spirometry (FEV1 or peak expiratory flow, PEF).
9. Administer two puffs of salbutamol or the subject's prescribed bronchodilator. Ten minutes after bronchodilator administration, perform two peak flow maneuvers for PEF or spirometry for FEV1. Repeat any maneuvers that do not appear to be technically acceptable. Average two (FEV1) or two (PEF) technically acceptable readings, and record as the pre-sputum induction baseline. Calculate 80% and 90% of the pre-sputum induction baseline PEF or FEV1 and record.
10. Record the start time on the worksheet, and start countdown timer (approximately 10 min).
11. Insert 3 unit doses (of 4 mL each) of 3% hypertonic saline, (or up to 7% as discussed on a case-by-case basis) (total 12 mL) in the nebulizer cup if using an ultrasonic nebulizer, or 2 unit doses (total 8 mL) if using a jet nebulizer.
12. Start nebulization.
13. Instruct participant to breathe tidal (relaxed) breathing (with deeper inspirations as needed).
14. When the timer indicates the end of the (approximate) 10-min interval, stop the timer and stop the nebulizer.
15. Instruct the participant to spit all saliva into the marked saliva container.
16. Instruct the participant to huff a maximum of three times, cough, then expectorate any sputum directly into the designated sputum collection container.

    If the subject is symptomatic at the end of nebulization, proceed to the procedure listed in **step 21**.
17. If after completing nebulization, there are no such symptoms, measure PEF or FEV1. Evaluate the result, and proceed as follows:

    If the PEF or FEV1 is ≥80% of the pre-induction baseline, record the value on the worksheet, and continue the sputum induction procedure.

    If the PEF or FEV1 is <80% of the pre-induction baseline, perform another forced expiratory maneuver.

    If both values remain <80%, then record the best value on the worksheet, stop the sputum induction procedure, and proceed to **step 21**.

18. If the participant is tolerating the procedure well, perform up to four additional cycles of **step 12** (if necessary, check the volume of the remaining hypertonic saline), **steps 13–15, 17**, and **18**.
19. If the last recorded FEV1 is >90% of baseline, discharge the participant.

    If it is below 90%, proceed to **step 21**.
20. Sputum induction should be discontinued.
    (a) If the PEF / FEV1 value at the completion of the procedure is <90% of the pre-induction baseline.
    (b) If oxygen saturation decreases by 5% or more from baseline level.
    (c) If oxygen saturation drops below 90% (on a case-by-case basis).
    (d) If participant feels unwell or has severe coughing.
21. Administer two puffs of salbutamol (100 μg/dose).
22. Ten minutes after the last puff, measure PEF/FEV_1. If the PEF/FEV1 is ≥90% of the pre-induction baseline peak flow and oxygen saturation has returned to baseline, proceed to discharge the participant.
23. Conversely, if the PEF/FEV1 remains <90% or oxygen saturation has not returned to baseline, the information should be reported immediately to the supervising physician. The participant should not be discharged without the approval of the supervising physician.
24. The collected sputum container should be stored at −80 °C.

3.2 Isolation of Airway Mucin from Induced Sputum

1. Homogenize 16 mL induced sputum with 80 mL ice-cold isotonic saline in a beaker by multiple aspirations with a 10 mL sterile plastic syringe (0.9 × 40 mm cannula).
2. Add dropwise 1 mL 1%-sodium azide to inactivate microbes.
3. Stir (10–20 rpm) homogenate at 4 °C for 24 h.
4. Centrifuge homogenate at 2000 g for 10 min.
5. Collect supernatant.
6. Wash the residue with 50 mL ice-cold isotonic saline.
7. Centrifuge homogenate at 2000 g for 10 min.
8. Collect supernatant.
9. Wash the residue with 50 mL ice-cold isotonic saline.
10. Centrifuge homogenate at 2000 g for 10 min.
11. Collect supernatant.
12. Combine the supernatants, and add 0.1 M EDTA to a final concentration of 1 mM EDTA.

13. Dialyze for in total 72 h at 4 °C against two changes of 0.01 M NaCl and four changes of bidistilled water.
14. Freeze-dry.
15. Dissolve the dry lyophilisate in 40 mL 0.5 M Tris-HCl, pH 8.0.
16. Add slowly under constant stirring 20 drops of citraconic anhydride.
17. Dialyze for in total 24 h at 4 °C against two changes of bidistilled water.
18. Freeze-dry.
19. Dissolve the dry lyophilisate in 20 mL bidistilled water, and adjust under stirring with acetic acid to pH 4.8.
20. Precipitate the mucin by slowly adding dropwise a total of 0.5 mL 20% (w/v) N-cetyl-N,N,N trimethyl ammonium bromide.
21. Keep the solution for 12 h at 4 °C.
22. Centrifuge for 30 min at 50,000 g at 4 °C.
23. Dissolve the sedimented mucin in a small volume (10–15 mL) 0.5 M NaCl.
24. Dialyze for in total 24 h at 4 °C against two changes of bidistilled water.
25. Dilute the dialysate in 150 mL 0.5 M NaCl.
26. Dialyze for in total 24 h at 4 °C against two changes of bidistilled water.
27. Freeze-dry (*see* **Note 7**).
28. Determine yield (± 0.1 mg) with an analytical balance.

3.3 Bacterial Growth

3.3.1 Preparation of Preculture

1. Streak an inoculation loop of the *P. aeruginosa* isolate stored in the micro-tube onto a LB agar plate (*see* **Note 8**).
2. Incubate overnight at 37 °C (*see* **Note 9**).

3.3.2 Checking Bacterial Growth Under Standard Conditions

1. Inoculate 25 mL LB in a 100 mL culture flask with a colony from the agar plate.
2. Cultivate overnight at 150 rpm at 37 °C.
3. The next morning inoculate fresh 25 mL LB medium with 1 mL of the overnight culture.
4. Cultivate at 150 rpm and 37 °C for 5 h.
5. Determine the OD_{600} of an aliquot of 1 mL at the spectrophotometer.
6. Dilute an aliquot of 2 mL with PBS to OD_{600} of 0.6.
7. Aspirate 9 mL PBS with a sterile glass pipette, and then aspirate 1 mL of the bacterial suspension (*see* **Note 10**).

8. Perform serial tenfold dilutions with PBS with sterile 10 mL glass pipettes.
9. Plate 0.1 mL aliquots of 10^5, 10^6, and 10^7 dilutions with a Drigalski spatula in triplicate.
10. Incubate overnight, and count colonies. Determine the CFU equivalent to OD_{600}.

3.3.3 Checking Bacterial Growth Under Nonstandard Conditions

1. Inoculate 25 mL medium in a 100 mL culture flask with a colony from the agar plate.
2. Cultivate at 150 rpm at 37 °C until the suspension reaches an OD_{600} of at least 1.5.
3. Inoculate 8 mL into fresh 200 ml medium in a 1 l culture flask.
4. Cultivate at 150 rpm and 37 °C, and take every 2 h 1 mL aliquots to determine OD_{600} and CFU (see **steps 5–10** in Subheading 3.3.2) until the culture reaches the stationary phase (*see* **Note 11**).

3.4 Mucin Adhesion Test

3.4.1 Coating of Microplate with Mucin (see Note 12)

1. Forty-eight wells of the microplate are coated with 200 μL mucin solution per well and the other 48 wells with 200 μL PBS per well. Subsequently, the plate is kept at 37 °C for 18 h until the adhesion assay is started (*see* **Note 13**).

3.4.2 Adhesion to a Single Mucin

1. Follow **steps 1** and **2** of Subheading 3.3.1 and then **steps 1–6** of Subheading 3.3.2.
2. Remove PBS from the mucin-coated microplate by suction (see Subheading 2.6).
3. Dilute the bacterial suspensions of OD_{600} 0.6 tenfold and hundred-fold with PBS.
4. Inoculate mucin-coated und uncoated wells of the microplate each in triplicate with either 100 μL of the tenfold or 100 μL of the hundred-fold dilution of the suspension of the bacterial strain.
5. Incubate the microplate (**step 5**) and the diluted bacterial suspensions (**step 1**) for 30 min at 37 °C.
6. Store the bacterial suspensions at 4 °C.
7. Remove the fluid from all 96 wells of the microplate with a Pasteur pipet by suction.
8. Wash all 96 wells 10 (!) times with 250 μL PBS per well. Use a multichannel pipet for adding PBS and remove PBS by suction with a Pasteur pipet (*see* **Note 14**).

9. Add 300 μL 1% Triton X 100 to each well and incubate for 30 min at room temperature to release mucin and bound bacteria.
10. Determine CFU by plating tenfold dilutions of a 100 μL aliquot (*see* **Note 15**).
11. Determine the CFU of the bacterial suspensions kept at 4 °C by plating tenfold dilutions of a 100 μL aliquot (*see* **Note 16**).
12. Incubate plates overnight and count colonies. The read-out is the ratio of mucin-bound bacteria to total number of bacteria.

3.4.3 Competition Experiments with Mono- or Oligosaccharides

Follow the same procedure as outlined above (*see* **Note 17**) in Subheading 3.4.2 with the exception of **step 5**:

Instead, add to all wells 50 μL of a 300 isoosmolar monosaccharide solution or 50 μL of an oligosaccharide solution (0.1–50 μg/mL), and then inoculate mucin-coated and uncoated wells of the microplate each in triplicate with either 100 μL of the tenfold or 100 μL of the hundred-fold dilution of the suspension of the bacterial strain.

4 Notes

1. The binding assay described here should be generally applicable to studies on the binding of Gram-negative bacteria to submaxillary or airway mucins, although some modifications may be necessary. More optimization will be necessary to adapt the protocol to adhesion studies with Archaea or Gram-positive bacteria, particularly actinobacteria-like mycobacteria and nocardia, which carry rigid acid-resistant cell walls.
2. Monosaccharides can be charged or uncharged. To allow an unbiased comparison between monosaccharides in competition experiments, isoosmolar sugar concentrations should be used. Since the affinity is rather low, a concentration of 150 isoosmolar is suggested for the first screen.
3. *O*-linked oligosaccharides may be synthesized upon demand from an academic laboratory specialized in the organic chemistry of carbohydrates. Recent publications [33] demonstrate the diversity of hundreds of *O*-glycans. Suggestion: At least the two common disaccharides Gal ß1–3 GalNAc—*O*- R and GlcNAc ß1–3 GalNAc—*O*- R may be compared.
4. Ovine submaxillary mucin (OSM) and asialo ovine submaxillary mucin are highly suited reference *O*-glycoproteins for bacterial binding assays to mucin because essentially all possible serine/threonine residues are occupied with homogeneous *O*-glycans, i.e., NeuAcα2-6GalNAcα1-O-Ser/Thr and 6GalNAcα1-O-Ser/Th, respectively [31].

Porcine submaxillary mucin (PSM) carries eight different *O*-glycan structures in its highly glycosylated tandem repeats, i.e., a α-GalNAc-O-Ser/Thr monosaccharide, a ß-Gal(1--3)-α-GalNAc-O-Ser/Thr disaccharide, a α-Fuc(1–2)-ß-Gal (1–3)-α-GalNAc-O-Ser/Thr trisaccharide, or a [α-GalNAc (1–3);α-Fuc(1, 2)]-ß-Gal(1–3)-α-GalNAc-O-Ser/Thr tetrasaccharide [32]. For each oligosaccharide a NeuNGl, sialic acid, residue may be 2,6-linked to the peptide core GalNAc residue.

Bovine submaxillary mucin (BSM) is commercially available; however, its *O*-glycans are diverse impeding any quantitative assignment of bacterial binding to a specific glycan. A recently published LC-MS analysis identified 275 *O*-glycans in a BSM sample [33]. The majority of structures contained sialic acids (Neu5Ac or Neu5Gc) often modified by mono-, di-, or triacetylation. Moreover, fucosylated and sulfated structures devoid of sialic acid were observed.

5. *P. aeruginosa* may be either collected from the habitat of interest or may be obtained from a reference strain collection. Since the protocol focuses on the interaction of *P. aeruginosa* with human airway mucin, it makes sense to retrieve *P. aeruginosa* strains from *P. aeruginosa*-positive human airways by deep throat swab, induced sputum, or spontaneously expectorated sputum. Samples should be processed by a proficient clinical microbiology laboratory. Swab or sputum is plated onto MacConkey, Mueller-Hinton, tryptic soy, and chocolate agar supplemented with 8% erythrocytes and 2% horse serum. *P. aeruginosa* colonies are discerned by the production of pigments (pyocyanine and pyoverdine), metallic gloss, and growth at 42 °C and then identified by MALDI-TOF MS (matrix-assisted laser desorption ionization–time of flight mass spectrometry) [34]. Alternatively, one may ask for isolates of publicly available *P. aeruginosa* strain collections, for example, the major *P. aeruginosa* clone type (mPact) panel from Hinrich Schulenburg, Kiel [35], or the two international *P. aeruginosa* reference strain panels from Joanne Fothergill, Liverpool [36], and Roger Levesque, Quebec City [37]. Moreover, one may ask the head of laboratories for *P. aeruginosa* strains who submitted the complete genome sequence of these strains to the Pseudomonas Genome Database [38].

6. Prior to starting any collection of induced sputum from volunteers, the operator should be trained by physiotherapists who are proficient in the inhalation therapy of people with chronic lung diseases. Sampling should be performed in a clinical setting that can provide supplemental oxygen or bronchodilation for people at risk. A supervising physician with expertise in pneumology should be available in-house for consultation if

during the collection of induced sputum, the PEF/FEV1 remains <90% below baseline or oxygen saturation has not returned to baseline (see **steps 21** and **24** of Subheading 3.1.)

According to the author's experience, induced sputum was successfully collected from more than 90% of healthy volunteers and people with COPD, bronchiectasis, or CF.

Contraindications. Induced sputum is prohibited to be collected from subjects suffering from:

(a) Hemoptysis.

(b) Acute respiratory distress.

(c) Unstable cardiovascular status (arrhythmias, angina).

(d) Thoracic or cerebral aneurysms.

(e) Hypoxia (SaO2 less than 90% on room air).

(f) Lung function impairment (FEV1 less than 1.0 liter).

(g) Pneumothorax.

(h) Pulmonary emboli.

(i) Fractured ribs or other chest trauma.

(j) Recent eye surgery.

(k) Unstable asthma.

(l) Low platelets.

(m) Severe bronchospasm.

(n) Large pleural effusions.

Caution should be exercised with participants who have previously been intolerant to hypertonic saline or who have demonstrated a constriction of >10% of their FEV1 from baseline 15 min posttest dose.

Detailed up-to-date Standard Operating Procedures (SOPs) for sputum induction are not freely available to the public but may be obtained upon request from the learned societies like the European Cystic Fibrosis Society (ECFS), European Respiratory Society (ERS), or American Thoracic Society (ATS). For example, the ECFS Clinical Trials Network issued a SOP by January 31, 2024: Sputum induction for expectorating children and adults, V2.0.

7. Depending on the resources and equipment of the site, the molecular weight and polydispersity of the mucin may be characterized by analytical ultracentrifugation or gel filtration. The absence of low-molecular-weight proteins may be checked by 10% SDS-PAGE. The gross sugar composition can be determined by thin layer chromatography (TLC), gas chromatography (GC), or GC/MS.

8. The mucin binding assay has only a small window for the CFU exposed to mucin-coated wells in a microtiter plate. Hence, the

number of CFUs exposed to mucin should be reliably estimated at the start of the experiment. Under standard conditions of growth in nutrient-rich medium in vitro, the density of a *P. aeruginosa* suspension in midexponential phase of an OD at 600 nm of 0.6 is equivalent to 10^9 CFU/mL. However, during the late exponential phase or stationary phase, the production of pigments and/or exopolysaccharides will distort this relation. Likewise, growth in minimal medium will generate other growth characteristics. Thus, if numerous strains are compared in their adhesion to mucin, each selected strain should be grown to exponential phase in LB, and the equivalence of an OD_{600} of O.6 with a CFU of 10^9 CFU/mL should be checked by spectrophotometry and plating prior to execution of the mucin binding assay. Moreover, if the bacterial strain is grown in media poor in nutrients or deficient in peculiar substrates or if the bacteria are harvested from lag, late exponential, or stationary phase, the strain-specific growth should be recorded by spectrophotometry and CFU determinations.

9. The agar plate may be stored at 4 °C for up to 10 days.
10. Dilutions of bacterial suspensions only work reliably if old-fashioned tenfold dilutions are performed with glass pipettes at the macroscopic 10 mL scale. If the work program uses repeatedly the same set of strains of known adhesion features to plastic, glass pipettes may be substituted by pipette tips and micro-pipettes.
11. Workload may be lowered if the work program specifies a growth phase for the mucin adhesion experiments.
12. Initially, a pilot binding assay may be run just on mucin binding in duplicate without negative control to get an initial clue whether the *P. aeruginosa* strain is a poor or strong mucin binder. This knowledge may guide the dilutions to be plated for the determination of CFUs.
13. To ensure reproducibility of mucin adhesion assays performed on different microplates, one should perform an adhesion study with the same lot of microplates. Nonspecific binding of *P. aeruginosa* to microplates differs by polymer and manufacturer and is a strain-specific feature. According to the author's laboratory experience, the unmodified polystyrol is a more reproducible negative control that siliconizes the surface of the well with a reagent such as Sigmacote.
14. If available, a well washer may be used for this step.
15. You should detect at least 1% binding of bacteria. Thus, plating 100 μL of 10^2-, 10^3-, and 10^4-fold dilutions should cover the whole range.

16. The final workload of **steps 11** and **12** may be high. Thus, it is advisable that a team of two to five dedicated people perform the CFU determinations. Agar plates and sterile pipette tips should be available beforehand in sufficient amounts. If possible by manpower, the input of bacteria may already be processed by plating the OD_{600} dilutions parallel to the washing step.
17. Oligosaccharides are scarce and still a precious resource. Competition experiments should be confined to the strongest binders. Comparison of bacterial adhesion to submaxillary mucins and to airway mucins from healthy donors and donors with chronic lung disease will already provide ample information about the broad diversity of mucin adhesion of *P. aeruginosa* strains.

References

1. McNamara NA, Sack RA, Fleiszig MJ (2000) Mucin—bacterial binding assays. Meth Mol Biol 125:429–437
2. Fahy JV, Dickey BF (2020) Airway mucus function and dysfunction. N Engl J Med 363: 2233–2247. https://doi.org/10.1056/NEJMra0910061
3. Lai SK, Wang YY, Wirtz D et al (2009) Micro- and macrorheology of mucus. Adv Drug Deliv Rev 61:86–100. https://doi.org/10.1016/j.addr.2008.09.012
4. Abrami M, Biasin A, Tescione F et al (2024) Mucus structure, viscoelastic properties, and composition in chronic respiratory diseases. Int J Mol Sci 25:1933. https://doi.org/10.3390/ijms25031933
5. Hansson GC (2020) Mucins and the microbiome. Annu Rev Biochem 89:769–793. https://doi.org/10.1146/annurev-biochem-011520-105053
6. Ermund A, Trillo-Muyo S, Hansson GC (2018) Assembly, release, and transport of airway mucins in pigs and humans. Ann Am Thorac Soc 15(Suppl 3):S159–S163. https://doi.org/10.1513/AnnalsATS.201804-238AW
7. Pust MM, Wiehlmann L, Davenport C et al (2020) The human respiratory tract microbial community structures in healthy and cystic fibrosis infants. NPJ Biofilms Microbiomes 6: 61. https://doi.org/10.1038/s41522-020-00171-7
8. Pienkowska K, Pust MM, Gessner M et al (2023) The cystic fibrosis upper and lower airway metagenome. Microbiol Spectr 11: e0363322. https://doi.org/10.1128/spectrum.03633-22
9. Roy MG, Livraghi-Butrico A, Fletcher AA et al (2014) Muc5b is required for airway defence. Nature 505:412–416. https://doi.org/10.1038/nature12807
10. Ermund A, Meiss LN, Rodriguez-Pineiro AM et al (2017) The normal trachea is cleaned by MUC5B mucin bundles from the submucosal glands coated with the MUC5AC mucin. Biochem Biophys Res Commun 492:331–337. https://doi.org/10.1016/j.bbrc.2017.08.113
11. Ermund A, Meiss LN, Dolan B et al (2018) The mucus bundles responsible for airway cleaning are retained in cystic fibrosis and by cholinergic stimulation. Eur Respir J 52: 1800457. https://doi.org/10.1183/13993003.00457-2018
12. Bos MF, Ermund A, Hansson GC et al (2023) Goblet cell interactions reorient bundled mucus strands for efficient airway clearance. PNAS Nexus 2:pgad388. https://doi.org/10.1093/pnasnexus/pgad388
13. Fernández-Blanco JA, Fakih D, Arike L et al (2018) Attached stratified mucus separates bacteria from the epithelial cells in COPD lungs. JCI Insight 3:e120994. https://doi.org/10.1172/jci.insight.120994
14. Ma J, Rubin BK, Voynow JA (2018) Mucins, mucus, and goblet cells. Chest 154:169–176. https://doi.org/10.1016/j.chest.2017.11.008
15. Padra M, Benktander J, Padra JT et al (2021) Mucin binding to Moraxella catarrhalis during airway inflammation is dependent on sialic

acid. Am J Respir Cell Mol Biol 65:593–602. https://doi.org/10.1165/rcmb.2021-0064OC

16. Burcham LR, Bath JR, Werlang CA et al (2022) Role of MUC5B during group B streptococcal vaginal colonization. MBio 13: e0003922. https://doi.org/10.1128/mbio.00039-22
17. Tümmler B (2022) What makes Pseudomonas aeruginosa a pathogen? Adv Exp Med Biol 1386:283–301. https://doi.org/10.1007/978-3-031-08491-1_10
18. Botelho J, Tüffers L, Fuss J et al (2023) Phylogroup-specific variation shapes the clustering of antimicrobial resistance genes and defence systems across regions of genome plasticity in Pseudomonas aeruginosa. EBioMedicine 90:104532. https://doi.org/10.1016/j.ebiom.2023.104532
19. Jolley KA, Bray JE, Maiden MCJ (2018) Open-access bacterial population genomics: BIGSdb software, the PubMLST.Org website and their applications. Wellcome Open Res 3:124. https://doi.org/10.12688/wellcomeopenres.14826.1
20. Lieleg O, Lieleg C, Bloom J et al (2012) Mucin biopolymers as broad-spectrum antiviral agents. Biomacromolecules 13:1724–1732. https://doi.org/10.1021/bm3001292
21. Kocevar-Nared J, Kristl J, Smid-Korbar J (1997) Comparative rheological investigation of crude gastric mucin and natural gastric mucus. Biomaterials 18:677–681. https://doi.org/10.1016/s0142-9612(96)00180-9
22. Schömig VJ, Käsdorf BT, Scholz C et al (2016) An optimized purification process for porcine gastric mucin with preservation of its native functional properties. RSC Adv 6:44932. https://doi.org/10.1039/c6ra07424c
23. Snyder CE, Nadziejko CE, Herp A (1982) Isolation of bronchial mucins from cystic fibrosis sputum by use of citraconic anhydride. Carbohydr Res 105:87–93. https://doi.org/10.1016/s0008-6215(00)81856-x
24. Radicioni G, Kesimer M (2024) Quantitation of MUC5AC and MUC5B by stable isotope labeling mass spectrometry. Methods Mol Biol 2763:125–136. https://doi.org/10.1007/978-1-0716-3670-1_11
25. Global Initiative for Chronic Obstructive Lung Disease (GOLD) (2023) Global strategy for prevention, diagnosis and management of COPD: 2023 Report. https://goldcopd.org/2023-gold-report-2/
26. Agustí A, Celli BR, Criner GJ et al (2023) Global initiative for chronic obstructive lung disease 2023 report: GOLD executive summary. Eur Respir J 61:2300239. https://doi.org/10.1183/13993003.00239-2023
27. Chalmers JD, Polverino E, Crichton ML et al (2023) Bronchiectasis in Europe: data on disease characteristics from the European bronchiectasis registry (EMBARC). Lancet Respir Med 11:637–649. https://doi.org/10.1016/S2213-2600(23)00093-0
28. Chalmers JD, Goeminne P, Aliberti S et al (2014) The bronchiectasis severity index. An international derivation and validation study. Am J Respir Crit Care Med 189:576–585. https://doi.org/10.1164/rccm.201309-1575OC
29. Grasemann H, Ratjen F (2023) Cystic fibrosis. N Engl J Med 389:1693–1707. https://doi.org/10.1056/NEJMra2216474
30. Boëlle PY, Viviani L, Busson PF et al (2012) Reference percentiles for FEV(1) and BMI in European children and adults with cystic fibrosis. Orphanet J Rare Dis 7:64. https://doi.org/10.1186/1750-1172-7-64
31. Konstantinidi A, Nason R, Čaval T et al (2022) Exploring the glycosylation of mucins by use of O-glycodomain reporters recombinantly expressed in glycoengineered HEK293 cells. J Biol Chem 298:101784. https://doi.org/10.1016/j.jbc.2022.101784
32. Gerken TA, Gilmore M, Zhang J (2002) Determination of the site-specific oligosaccharide distribution of the O-glycans attached to the porcine submaxillary mucin tandem repeat. Further evidence for the modulation of O-glycans side chain structures by peptide sequence. J Biol Chem 277:7736–7751. https://doi.org/10.1074/jbc.M111690200
33. Vos GM, Weber J, Sweet IR et al (2023) Oxidative release of O-Glycans under neutral conditions for analysis of Glycoconjugates having base-sensitive substituents. Anal Chem 95: 8825–8833. https://doi.org/10.1021/acs.analchem.3c00127
34. Hogardt M, Kahl BC, Besier S et al (2019) MIQ 24: Atemwegsinfektionen bei Mukoviszidose, 2nd edn. Elsevier
35. Tueffers L, Batra A, Zimmermann J et al (2024) Variation in the response to antibiotics and life-history across the major Pseudomonas aeruginosa clone type (mPact) panel. Microbiol Spectr 12:e0014324. https://doi.org/10.1128/spectrum.00143-24
36. De Soyza A, Hall AJ, Mahenthiralingam E et al (2013) Developing an international Pseudomonas aeruginosa reference panel. Microbiology 2:1010–1023. https://doi.org/10.1002/mbo3.141

37. Freschi L, Bertelli C, Jeukens J et al (2018) Genomic characterisation of an international Pseudomonas aeruginosa reference panel indicates that the two major groups draw upon distinct mobile gene pools. FEMS Microbiol Lett 365(14). https://doi.org/10.1093/femsle/fny120

38. Winsor GL, Griffiths EJ, Lo R et al (2016) Enhanced annotations and features for comparing thousands of pseudomonas genomes in the pseudomonas genome database. Nucleic Acids Res 44(D1):D646–D653. https://doi.org/10.1093/nar/gkv1227

INDEX

Inka Brockhausen (ed.), *Dynamics of Bacteria-Mucus Interactions*, Methods in Molecular Biology, vol. 2942, https://doi.org/10.1007/978-1-0716-4627-4,

O

P

Q

R

S

T

U

V

W

Z

www.ingramcontent.com/pod-product-compliance
Lightning Source LLC
Chambersburg PA
CBHW082203200625
28456CB00004B/53
9781071646267